# DEMYSTIFYING ORGANIC CHEMISTRY

## THE UDED METHOD

ROBERT J. TERNANSKY, Ph.D.

innovativeink
PUBLISHING
A Division of Kendall Hunt

Cover image Shutterstock.com

innovativeink PUBLISHING
A Division of Kendall Hunt

www.innovativeinkpublishing.com
*Send all inquiries to:*
4050 Westmark Drive
Dubuque, IA 52004-1840



Print ISBN: 979-8-3197-0016-2
eBook ISBN: 979-8-3197-0017-9



Published in the United States of America

# Student UDED Method Support Comments

"I took the MCAT this past summer and my CHEM/PHYS section was all organic chemistry and I surprisingly did the best in that section even though the only/main concept I applied was UDED."
**U.S.**

"Being introduced to your UDED philosophy has not only built the foundation for my intuitive understanding of mechanisms, but it has also given me a framework I would've taken years to intuitively understand on my own….as someone who struggles terribly with memory and relies heavily on conceptual understanding to grasp the driving forces behind reactions, I always find myself returning to the pillars of UDED.
**B.M.**

"Dr. Ternansky's UDED method helped me stop fearing chemistry and start enjoying it. It turned complex ideas into something I could actually reason through—and made me genuinely curious about the subject."
**K.B.**

"I wanted to say thank you because your class taught me the "UDED" philosophy which I have used in further Organic Chemistry courses and on the MCAT. I greatly appreciate this way of reasoning, and it has helped me approach complex problems with a clearer and more methodical mindset."
**A.T.**

“While studying for the Ochem section for the MCAT, I applied the UDED concept, which not only made everything easier, but it was the first time I actually felt like I knew what I was doing. I was understanding ochem. There’s a certain type of happiness people experience out there. But there’s nothing you can compare to the happiness I’m experiencing, now that I’m understanding Ochem through the concept you taught.
**E.S.**

“During my time in undergrad at UC Irvine, I had failed and retook every organic chemistry class with absolutely no hope of ever understanding this subject. It wasn’t until your class and UDED that for the first time, I understood and actually enjoyed what I was learning. I am currently studying for the MCAT, and because of your lessons I have been breezing through the organic chemistry section and I cannot begin express my gratitude. I am now even able to apply UDED to concepts reaching into biochemistry.”
**L.DeG.**

“I have taken your class for two quarters now and I want to express my appreciation for your concept of UDED that helped to significantly improve my understanding of organic chemistry.”
**M.L.**

“I recently took the MCAT and the organic chemistry portion of the MCAT turned out to be one of my best sections and I am pretty confident it was all due to UDED. I just wanted to thank you for teaching all your students UDED because its such a helpful way to understand organic chemistry!”
**R.V.**

“I want to sincerely thank you for your approach. It has sparked a LOVE for organic chemistry in me and inspired me to both change my major and the trajectory of my career goals. I wish my friends who failed/cried thru all of organic chemistry and came to loathe it had been introduced to UDED.“
**B.M.**

# CONTENTS

# PREFACE

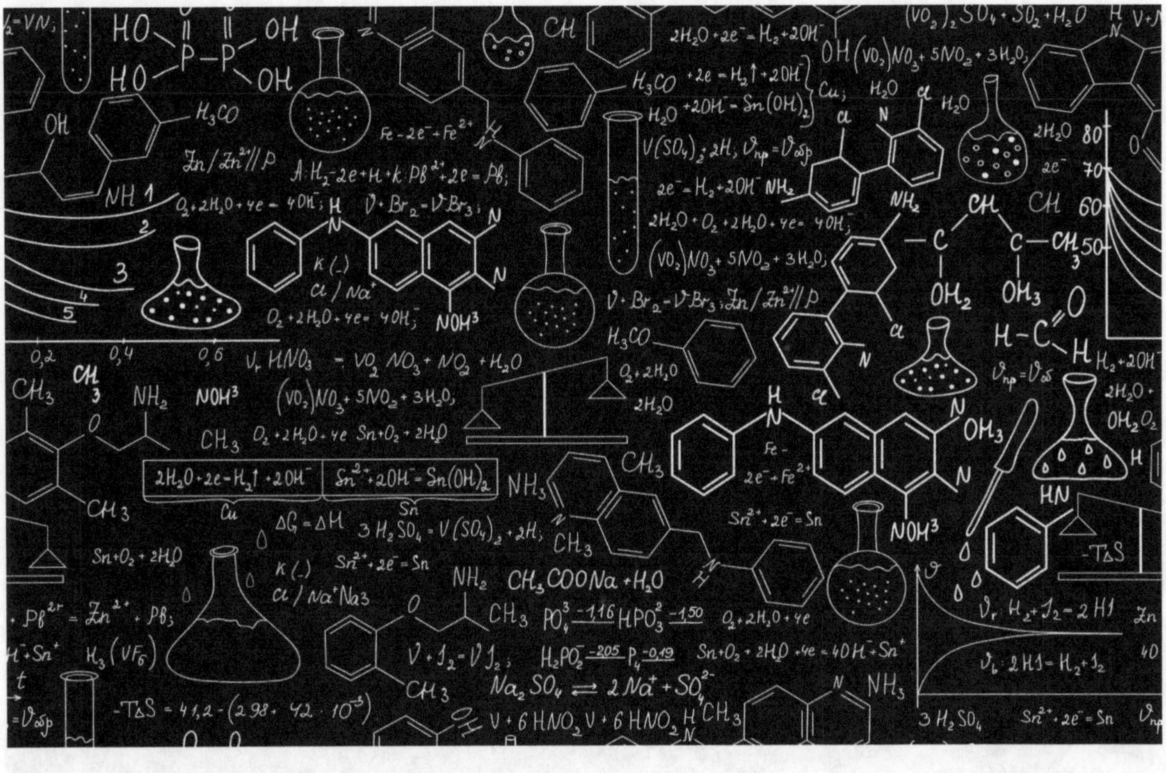

Marina Sun/Shutterstock.com

Organic chemistry, as a course of study, has a negative reputation. It is considered to be a very difficult subject to comprehend, and many consider it a "weed-out" class for those aspiring to careers in the medical field. As a required course of study for those majoring in

STEM fields, it is feared that, and as such, it is extremely rare to encounter a student who enrolls in an organic chemistry class as an elective.

Since I have dedicated my career to the field of organic chemistry, I find its reputation to be inappropriate and quite unnecessary. However, in my over a quarter century of teaching, I have come to realize why the subject is so confusing to those learning it for the first time. It is not because the average college student is not capable, but rather, in my opinion, it is how the topic is presented. In many cases, the topic of organic chemistry is presented as a dizzying array of random reactions, and the student, in desperation, sees the only survival method to be brute memorization without the advantage of a true understanding of why molecules actually behave as they do. Unfortunately, not only is this a monumental challenge due to the sheer volume of material, but once memorized, it is soon forgotten. The key to success is to truly understand the fundamental underlying concepts of organic chemistry and then to apply this understanding to the behavior of organic molecules. Memorization is not learning; it is fleeting, and although it may result in a passing grade for the course, it does nothing to enrich the knowledge or reasoning skills of the student. I believe that medical, dental, and pharmacy schools require a year of organic chemistry not to test a student's capability to memorize a huge volume of data (computers do a much better job of this and they don't forget!) or to weed people out but rather to train future medical professionals how to think logically, to reason, and to use these skills to solve new problems in the health fields.

paulista/Shutterstock.com

In order to help students of organic chemistry, I looked to unearth how my understanding of the subject evolved and to present it early in the learning process so that the seemingly disconnected behavior of organic molecules can be understood through a lens of common and simple patterns. Thus arose the "UDED" method, which provides a guide to effectively learning and understanding organic chemistry. "UDED" stands for the "Unequal Distribution of Electron Density," which, as you will see, is the cornerstone of organic molecular behavior. The UDED method of understanding organic chemistry has proven itself to be successful with many over the years, which has led me to continue to share the methods with as many students of the subject as possible. I hope that with this approach, students learning organic chemistry will find the subject much less intimidating and quite enjoyable.

Based on my experience in applying this method, I can confidently estimate that it will provide the learner with a clear understanding of 95-98% of organic chemistry. The other 2-5%, which is not readily explained by UDED, includes topics typically reserved for those continuing to an advanced chemistry level in graduate school.

I wish to make a very important point before moving on. None of the chemical concepts I present are new. They are well known, and you have most likely been introduced to these in the context of a general chemistry class. Also, there are many other sources that use the concept of differences in electron densities to guide the understanding of organic chemistry. What is different with the UDED approach is that it is a unifying theory. It (1) places the three most important ways of electron distribution in the same "toolbox" and (2) provides the student with a way to understand electron distribution by comparison with experiences in everyday life, making the topics easy to understand and apply to problem solving in organic chemistry.

Finally, this book is not meant to replace a traditional organic chemistry curriculum but rather to complement it by demystifying the many concepts presented during the presentation of the topic.

I would like to thank my many past students who have worked with the method and helped me to refine it to the level presented in this treatise.

# INTRODUCTION

One of the challenges faced by students of organic chemistry is that the underlying behavior of this class of molecules is not intuitive. In fact, there is little about the way molecules react that has been a part of our past life experiences. Other science-based subjects do not face a similar situation. For example, many of the sciences have a significant mathematical component underlying their behaviors. Students of these subjects have been exposed to basic mathematical concepts for most of their lives. Other sciences, such as biology, can easily be relatable by just living life and experiencing nature. However, organic chemistry does not present much that can be understood due to past experiences or innate human intuition.

At a thirty-thousand-foot level, understanding organic chemistry is best viewed as seeing the relationship between structure and function. In fact, this is the subtitle of an undergraduate textbook by Vollhardt and Shore ("Organic Chemistry Structure and Function", W.H. Freeman publisher). Intuitively, we can appreciate this relationship. For example, we can visualize the shapes of a hammer or a saw. Based on their structures, we know that we would take a hammer from our toolbox if we wanted to drive a nail or a saw if we wanted to cut a piece of wood. We also know that a hammer would not be a good choice if we wanted to cut something, nor a saw a good choice for driving a nail. In order to understand organic chemistry, these structure-function relationships that we are familiar with in our everyday lives (hammers and saws) need to be developed for organic molecules. Unfortunately, in many cases, the learning of organic chemistry goes straight to function without making the connection to molecular structure. The result is that the default "learning" method becomes memorization, which, as previously noted, is not a workable solution for this subject. This is where UDED comes in. Not only will the emphasis be on relating structure to function, but, more importantly, UDED will focus on the structural features that are the most important in directing chemical behavior. With practice, these UDED tools will become intuitive and will thereby allow us to readily understand and predict molecular behavior.

Soraya Plaithong/Shutterstock.com

zblaster/Shutterstock.com

Now, let's examine the topic of the structure of organic molecules. Unlike many familiar objects in everyday life, which only have a physical structure, organic molecules have both physical and electronic aspects to their structure. The physical characteristics of organic molecules include their size and shape. These aspects of molecular structure can be fairly intuitive. This is because we understand aspects of size and shape from our everyday experiences. Most infants were provided with a toy that helped them to develop this intuition. Throughout our lives, we are often presented with issues of size and shape: will this new sofa I just bought fit in my car, or will I need to have it delivered? Will all of these items I need to take on a trip fit in my suitcase? And even though we cannot "see" organic molecules, we can use molecular models to accurately represent their physical characteristics.

Refox Photos/Shutterstock.com

Therefore, the physical attributes of organic molecules, often referred to as "steric effects" or simply "sterics", can be intuitively understood. Also, although more challenging, students can be somewhat comfortable with the concept of stereochemistry from their experience of gazing into a mirror.

The problem is that, although very important, most chemical behavior is *not* driven by these familiar and intuitive themes of sterics or stereochemistry. Rather, the behavior of organic molecules is primarily driven by electronic effects. Unfortunately, most people do not have an intuitive concept of electricity or electron flow, or electron distribution, other than using electricity to power devices. But in reality, molecules mainly take action (react) when there is an imbalance in electronic distribution. And it is these actions that are the main focus for understanding organic chemistry. Therefore, in order to understand molecular behavior, it is imperative that we first become comfortable with electron density and electron flow. The good news is that this is not difficult to achieve. Developing this intuition is the basis of the "Unequal Distribution of Electron Density (UDED)" approach. This method is based on achieving a fundamental understanding of how electron density is distributed in organic molecules and how, when the electron density is unequally distributed, molecules will respond in an attempt to correct the imbalance.

## ASSUMED KNOWLEDGE

Since the UDED method is focused on electron density, one must first know the basics of electronic structure, such as the number of valence electrons per atom or ion; sigma and pi-bonding electrons; and non-bonding electrons. These topics are covered in general chemistry courses and will not be repeated here. Various drawings or depictions are employed to represent the electronic structure. I have provided a basic review of the common types of structural depictions used in organic chemistry in the Appendix. It is important that you become comfortable with these molecular representations before going further.

Once we know the number and location of electrons in a molecule, we need to understand how they can distribute throughout the molecule. Fundamentally, we need to appreciate the fact that not all electrons in a molecule are fixed in one location but can spread out. This distribution can occur throughout the entire molecule or it can take place in more localized and subtle ways. Understanding this distribution of electron density is the key to understanding molecular behavior, which is why it is the focus of the UDED method.

The distribution of electron density in molecules is achieved in three main ways: resonance, induction, and hybridization. Although these concepts are typically introduced in a general chemistry course, due to their importance to the UDED method, we will go over them in detail in this book. These are the main tools that will be present in our UDED toolbox, so it is essential that we learn how to use them properly.

# 1

# THE CULTURE AND SOCIETY OF ORGANIC MOLECULES

TANYARICO/Shutterstock.com

In order to better understand the dependency of molecular behavior on electron density, we first must understand the basic philosophy of the community of organic molecules. In order to do this, we will draw some analogies from our everyday life experiences.

First of all, it is important to realize that electrons are the "currency" of molecules. As with money in our daily lives, it provides molecules with stability and a means of obtaining necessary goods and services. But in the culture of organic molecules, attitudes regarding money (electrons) are somewhat different from what we as humans might experience. In the first place, although some atoms are more greedy than others, in general, molecular culture dictates that all members have, or at least strive to attain, equality of wealth. That is, molecules with excess wealth (electron density) are always looking for ways to share their wealth with others who experience a deficiency. The goal is a utopian molecular society in which no molecule is lacking in electron density and no molecules have an excess.

Another important attribute of molecular society, which derives from the first, is that opposites attract. Recall that electrons are negatively charged. Because of this and the fact that opposites attract, molecules with excess electron density (as represented by full or partial negative charges) are attracted to those with a deficiency of electron density (represented by full or partial positive charges). This is similar to the experience most of us have had when playing with magnets. That is, the positive (or negative) poles of two magnets will repel each other, whereas a positive pole of one magnet will attract the negative pole of another.

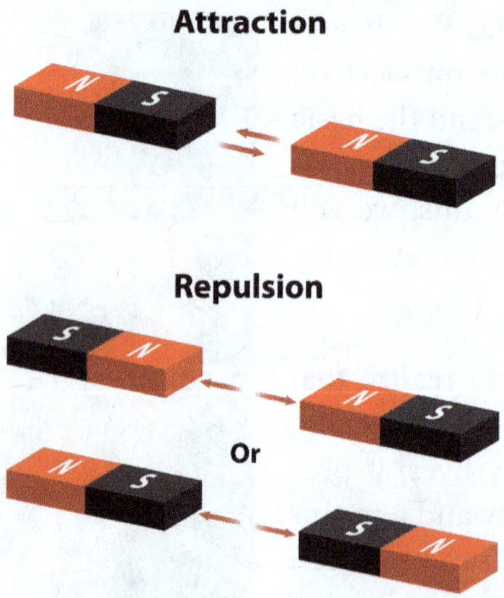


OSweetNature/Shutterstock.com

With this understanding and appreciation of the culture of organic molecules, we are now ready to examine the details of how electron density is distributed in molecules, and later we will apply this understanding to predict molecular behavior (the UDED method).

Before launching into the next section, it is important to note that there are no new chemical concepts being presented; they are all ones that you touched upon in general chemistry. The difference is that we will now gather them together into a single UDED "toolbox" and then learn how to use the tools to understand the seemingly disparate behavior of organic molecules. It will take some practice on your part, but once you learn the use of the tools in the UDED toolbox, you will discover that understanding 95-98% of the behavior of organic molecules is not only fairly straightforward but actually enjoyable!

# 2

# THE UDED METHOD

## THE EFFECTS OF ELECTRON DISTRIBUTION IN MOLECULES

The degree of electron distribution within a molecule will dictate its behavior. As previously stated, physical structure will also come into play, but it is secondary to electronic structure. Also, physical structure is easy to understand and apply, and as such is not a part of the UDED method. In general terms, in order to understand organic chemistry, it is most important to be able to relate electron distribution to the overall demeanor of molecules. That is, how does electron distribution affect a molecule's stability, reactivity, and selectivity? These attributes are interrelated and are directly influenced by the distribution of electron density. Let's discuss stability, reactivity, and selectivity, and see how they are interrelated.

Electron distribution will directly affect a molecule's stability. The more evenly distributed the electron density is within a molecule, the more stable it will be. We can look at stability as the molecule being "happy" or satisfied. Stability is achieved when each atom has an octet (or duet for hydrogens) and no atom is bearing excess electron density or a deficiency of electron density. The reactivity of a molecule is inversely related to its stability; the less stable, the more reactive. Conversely, the more stable a molecule is, the less likely it is to react. Finally, the more reactive a molecule is, the less selective it will be. That is, it will react with any other molecular entity it can find that will allow it to achieve a more balanced electron distribution.

We can relate a molecule's behavior to our own life experiences. Let's imagine that we had a busy day at school and didn't have time for lunch. Our bodies are experiencing a deficiency of calories. We are less stable and more reactive than normal, looking to find some food as soon as we can. And now, we find that we are much less selective with the food we consume; whatever is most readily available will suffice, even that leftover spaghetti! Molecules behave similarly with respect to their electron densities. Table 1 summarizes these relationships. Understanding these will allow us to predict the behavior of organic molecules.

TABLE 1

| DISTRIBUTION OF E-DENSITY | STABILITY | REACTIVITY | SELECTIVITY |
|---|---|---|---|
| MORE EQUAL | MORE | LESS | MORE |
| LESS EQUAL | LESS | MORE | LESS |

## HOW ELECTRON DENSITY IS DISTRIBUTED IN MOLECULES AND ITS EFFECTS ON STABILITY

As previously noted, there are three main ways that electron density is distributed within a molecule. These are (1) resonance, (2) induction, and (3) hybridization. Although hybridization is really a sub-category of induction, we will consider it separately. These basic chemical concepts should all be familiar to you from your general chemistry studies, but we will look at them in more detail here and, in particular, show how they relate to UDED. We will also assign them relative importance with respect to their significance in influencing chemical behavior.

Each of the three main modes of electron distribution (resonance, induction, and hybridization) can either provide (donate) electron density or remove (withdraw) it. We will see examples of both of these situations for each method as we go through them individually. But at this point, it is important to consider that, depending on the specific situation, the "direction" of electron distribution (providing or removing) can either be stabilizing or destabilizing. For example, if a molecule experiences an excess of electron density, withdrawing the electron density would prove beneficial and stabilizing. However, if additional electron density were to be provided to the molecule already burdened with an excess, the result would not be favorable. And of course, the opposite can be reasoned for a situation in which a molecule experiences a deficiency in electron density; electron donation will be good (stabilizing); electron withdrawal bad (destabilizing). Table 2 summarizes these situations. It is important to take a moment to be sure you understand these possibilities.

## TABLE 2

| UDED EFFECT | SITUATION | STABILIZING OR DESTABILIZATION |
|---|---|---|
| ELECTRON DONATING | LOW ELECTRON DENSITY | STABILIZING |
| ELECTRON DONATING | HIGH ELECTRON DENSITY | DESTABILIZING |
| ELECTRON WITHDRAWING | LOW ELECTRON DENSITY | DESTABILIZING |
| ELECTRON WITHDRAWING | HIGH ELECTRON DENSITY | STABILIZING |

Of course, in order to understand how electrons are distributed, one needs to know WHERE the electrons are located in a molecule. This requires the ability to construct accurate Lewis dot structures. This requires practice, and I recommend the book "Pushing Electrons" by Daniel Weeks (Brooks-Cole publisher), Chapter 1. There are also some practice problems provided at the end of this book. Be sure to become comfortable with creating Lewis structures before proceeding with the UDED method.

## DISTRIBUTION OF ELECTRON DENSITY VIA RESONANCE

Let's begin with the most important way electron density is distributed in molecules, and that is through resonance. In resonance, electrons are "mobile" and can flow through a connected system of unhybridized *p* orbitals. Such systems are referred to as "conjugated" systems. Resonance serves to distribute excess electron density over multiple atoms, thereby relieving the necessity of one atom to bear all the "weight" of the excess electrons while allowing other atoms with a deficiency of electrons to receive some of the desired density. Similarly, a deficiency of electron density can be shared among the atoms of the conjugated system. This resonance sharing is beneficial to the molecule because it provides for a more equal distribution of electron density. But resonance is not necessarily always beneficial, as we will see in later examples. That is, resonance can also lead to a more unequal distribution of electron density, leading to less stability.

It is important at this point to pause and explain what is meant by "excess" electron density. In the context of UDED, excess electron density means that the molecule can remain structurally intact without those "excess" electrons. That is, although the molecule certainly won't be the same without them, the overall molecule will remain intact; it will not have been broken apart. Said another way, electrons, which make up sigma bonds, are critical for the overall integrity of the molecule's structure; they are never considered to be "excess". However, electrons in pi-bonds or lone electrons or lone electron pairs are considered "excess" for the UDED method. Figure 1 provides an example.

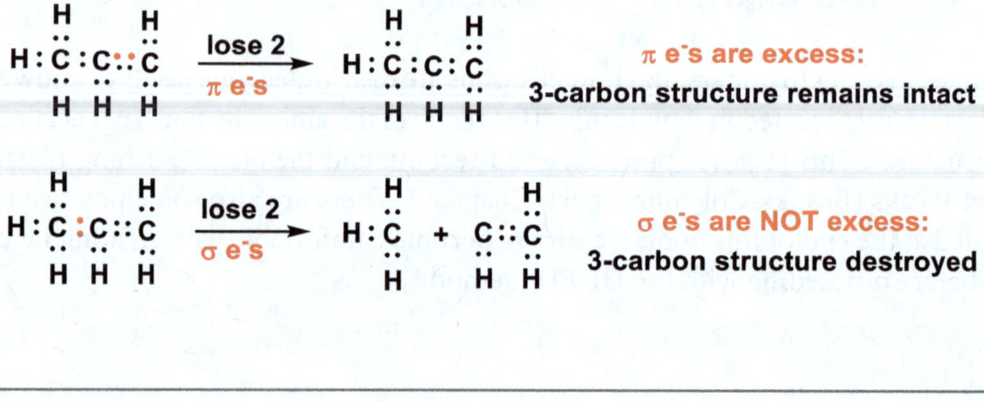


FIGURE 1

Returning to the distribution of electron density via resonance, it is important to be able to correctly draw resonance structures. It is also important to understand the relative quality of resonance contributors. This takes practice in drawing resonance structures. I find many students just want to look at a structure and imagine resonance contributors. Although this is possible after working on many examples, it will not be successful if one tries to skip the practice part. For practice, I recommend the book "Pushing Electrons" by Daniel Weeks (Brooks-Cole publisher), Chapter 2. There are also some practice problems provided at the end of this book.

Now, what is the effect of resonance delocalization on the distribution of electron density within a molecule? Or, more specifically, how can resonance increase (or decrease) the stability of a molecule that is experiencing an imbalance in electron density? We can compare the stabilizing effect of resonance to a common life experience of carrying a heavy object. Let's say that you and your friend go to the bookstore and purchase several textbooks. If, when taking the textbooks home, you must carry all the books, it can be a challenge. In fact, it can cause you to become "unstable". However, if your friend offers to carry some of the heavy books, the burden becomes distributed between two people, and no individual is overburdened. This is how resonance works beneficially; it distributes the burden among several atoms.

Roman Samborskyi/Shutterstock.com

stockfour/Shutterstock.com

With regard to resonance as a factor in the distribution of electron density within molecules, it is important to know that this is the most important of the three (the others being induction and hybridization). For this reason, when utilizing UDED, one should always be sure to include an evaluation of resonance effects. Sometimes, induction or hybridization effects will conflict with resonance, but in these cases, resonance will usually (95-98% of the time) be the dominating influence on electron density distribution.

Before continuing, you should be made aware of common terminology for carbon atoms bearing an unequal distribution of electron density in the form of full positive or negative charges. Carbons bearing a full negative charge are termed "carbanions," whereas those with a full positive charge are known as "carbocations".

Let's look at an example of two molecules bearing excess electron density (Figure 2). In the propyl anion (**1**), all the excess electron density is "carried" by one carbon atom; it is not very stable. However, in the propenyl anion (**2**), resonance allows the burden of the excess electron density to be shared between two carbon atoms. Molecule **2** is more stable (less reactive) than molecule **1**.

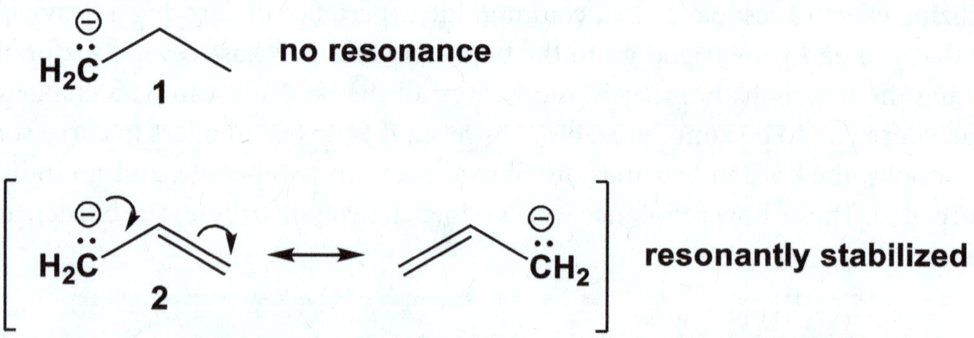


**FIGURE 2**

This is one reason that the carbon attached to a doubly bonded carbon is given a special name, the "allylic carbon," since it can be a stable place for unequal electron densities to exist.

Similarly, deficiencies in electron density can be stabilized by distributing the deficiency among other atoms (Figure 3).

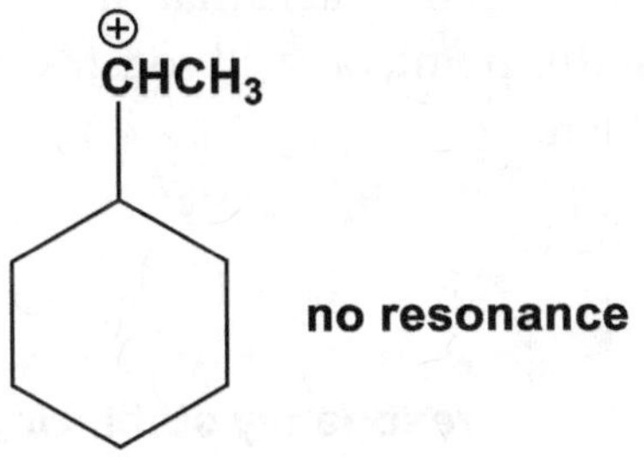


3

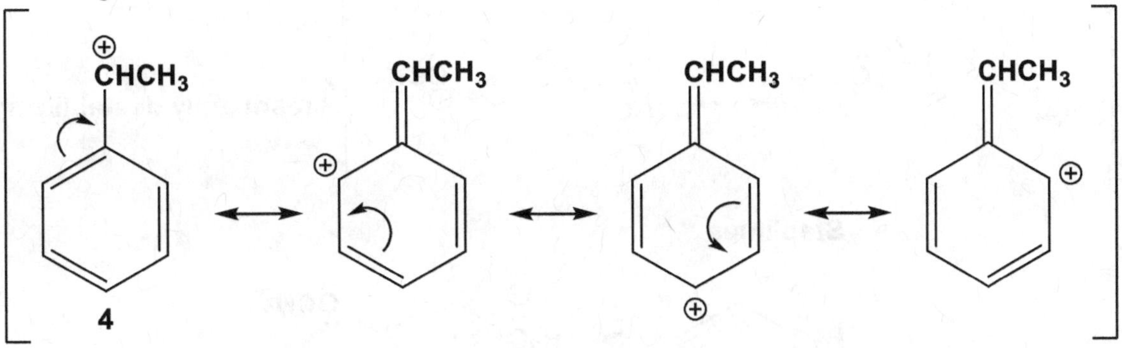


4

resonantly stabilized

FIGURE 3

Once again, the carbon directly attached to the ring in **4** is given a special name, the "benzylic carbon," since it can be a stable place for unequal electron densities to exist.

Now, as was stated previously, resonance is not always a stabilizing force within the distribution of unequal electron densities. For example, we can look at the examples in Figures 2 and 3 and now add substituents to the molecules, which will effectively *destabilize* the unequal distribution of electron density via resonance.

If we add a methoxy group to the carbanion **2** to form carbanion **5**, we can see that, in addition to the stabilizing effect of resonance, there now exists an opposing destabilizing resonant effect. In this situation, the question will arise as to which of the two effects will predominate. This can be predicted by assessing the quality of the resonance contributors (see Appendix). In this case, the resonance contributor having a formal positive charge localized on oxygen will be less favorable, albeit valid. However, the important aspect of this UDED evaluation is to compare the two molecules with and without the methoxy group. And, based on resonance, molecule **2** (Figure 2) is predicted to be more stable than molecule **5** (Figure 4). Note that we are only considering resonance at this point, but will consider other effects of electron distribution (induction/hybridization) later.

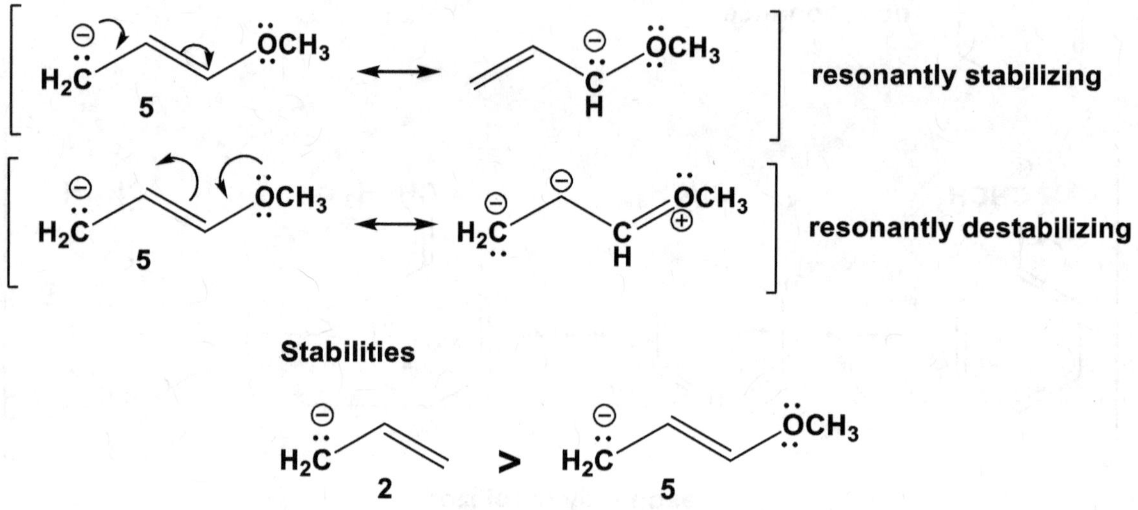


FIGURE 4

Can the carbocation **4** (Figure 3) be destabilized through resonance? Let's add a nitro group to the 6-membered ring and examine one resonance contributor. Although we have only drawn one of several possible resonance contributors to **6** (Figure 5), we can see that this contributor places another positive charge next to the original positive charge. This, of course, is destabilizing (remember the magnet example) and will result in the stability of compound **6** being less than that of **4**.

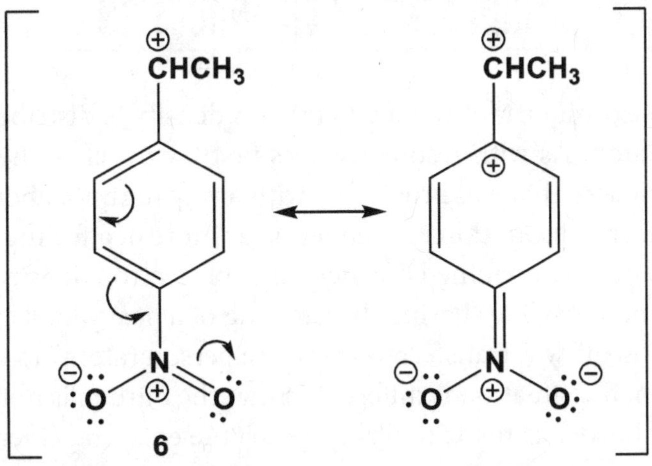


**Stabilities**

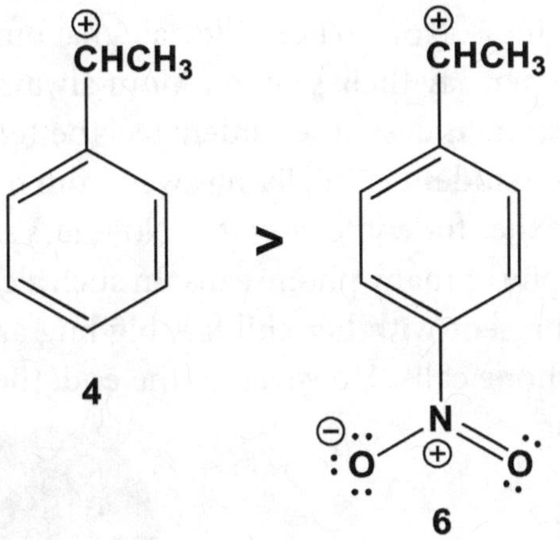


**FIGURE 5**

As a final note, it should be mentioned that although for the purposes of evaluating stabilities (and thereby reactivities) of molecules with UDED via resonance, although we can visualize this as electrons being able to "move" throughout the conjugated system, in reality, the electrons do not actually move but are distributed in a manner that best represents the hybrid structure (see appendix for further discussion of this point).

## DISTRIBUTION OF ELECTRON DENSITY VIA INDUCTION

Moving on to the next important way that electron density is distributed in molecules, we will consider induction. As with resonance, let's first try to relate the inductive effect to a common life experience. This will provide us with a way to think about induction that will be easier to remember. I like to think of induction as more of a feeling rather than an actual action. With resonance, we imagined the inequality of electron density being shared among more than one atom, much like sharing the carrying of a heavy load among several people. In these cases, the "load" was transferred from one person (atom) to other people (atoms). Now with induction, the situation is different in that the burden is not shared, but the individual carrying the burden is made to feel better by the encouragement of others.

Let's take an example from our everyday experiences. Many students who go to college experience living away from home for the first time. In these situations, the student often experiences the burden of loneliness. Now to help alleviate that burden, the student can call home and talk to a family member, say their mother. Mom always knows how to make her child feel better with her words, and indeed, the student feels better. But notice that the burden still exists with the student; the student is still living away from home, away from family. That reality has not changed, yet the student feels better. Now, let's say that the student still feels the burden of loneliness in spite of many phone calls. In such a case, mom might decide to travel to the college and be present with her child. A big hug and face-to-face interaction give much more relief than phone calls. However, in the end, the situation remains unchanged; the student is still living away from home.

PeopleImages/Shutterstock.com

Now, let's use this real-life example and relate it to the burden of the unequal distribution of electron density in molecules. In these cases, a molecule is experiencing a burden of too much or too little electron density. Now "mom" can be represented by an atom that is electron withdrawing or donating via induction. What is induction due to? It is due to the atom's predilection to withdraw or donate electron density via the phenomenon of electronegativity. Importantly, the effect of induction takes place through sigma bonds, not through unhybridized *p* orbitals as with resonance.

Let's look at some examples. In the figure below, we consider carbon bearing a negative charge (a carbanion). Carbon is not very electronegative and, as such, it is not very happy to be burdened with all of this excess electron density. However, if we attach an electron-withdrawing atom to the carbon (one with a higher electronegativity), the result will be to make the carbon more stable. The carbon feels better because a close, electronegative atom (chlorine in this example) is showing it can help with burden, albeit only "psychologically"; the negative charge still resides on the carbon, which did not change. But the presence of the electron-withdrawing chlorine makes the carbon more stable; it "feels" better. We represent this inductive effect with an arrow that indicates the direction of pull of the electron density (in this case, towards the chlorine atom). The cross at the tail end of the arrow indicates this position is losing electron density and developing a partial positive environment, which, of course, is stabilizing to the adjacent negative charge.

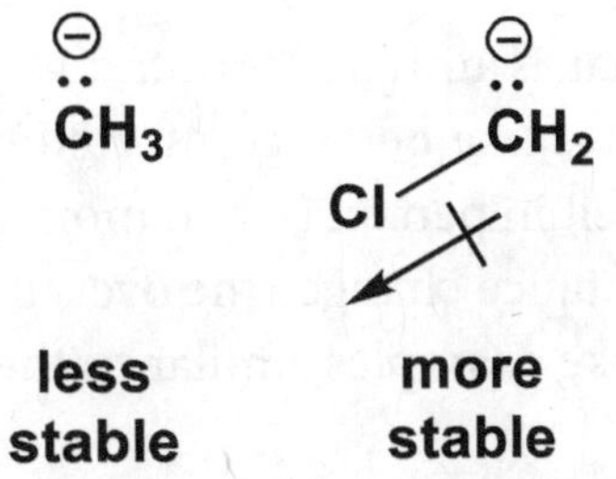


If we add additional electron-withdrawing atoms, we increase the stability even more. In Figure 6, we rank the stabilities of four carbanions based on the inductive withdrawing effect of chlorine.

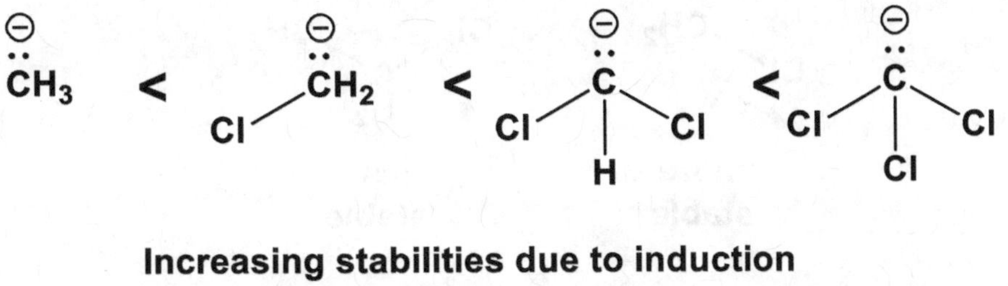


FIGURE 6

We should always keep in mind that the main ways in which electron density is distributed in molecules (resonance, induction, and hybridization) can produce stabilizing as well as destabilizing effects. Let's take the same molecules, but in place of an excess of electron density on carbon, we have a deficiency. In this case (Figure 7), adding electron-withdrawing groups to the positively charged carbon makes matters worse and decreases the stability. The electron-deficient carbon needs electron density donated to it for stability. The extra chlorine atoms enhance the deficiency of electron density, making the molecules less stable (more reactive).

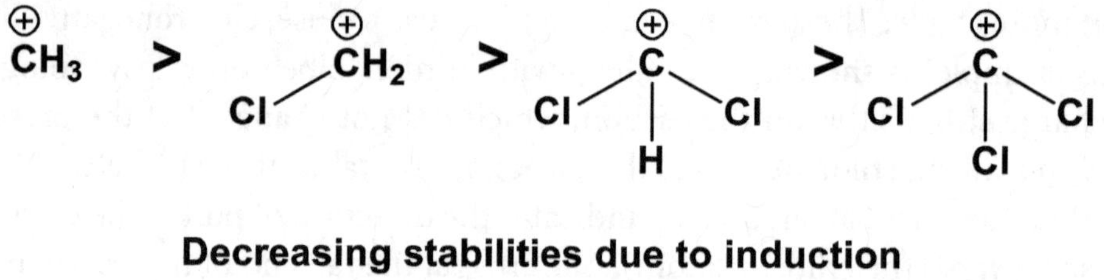


FIGURE 7

An important consideration that is unique to induction is that the effects will decrease significantly with distance. This is like being comforted by someone who is physically far away versus that same comfort felt in person (when mom comes to visit). Let's take a look at similar molecules to see how distance changes the overall impact of induction with respect to electron distribution. We'll take examples similar to the ones we have just examined.

In the case shown in Figure 8 below, we now move the electron-withdrawing (and stabilizing) chlorine atom further away from the site of unequal electron density. The further away the chlorine moves, the less its stabilizing effect.

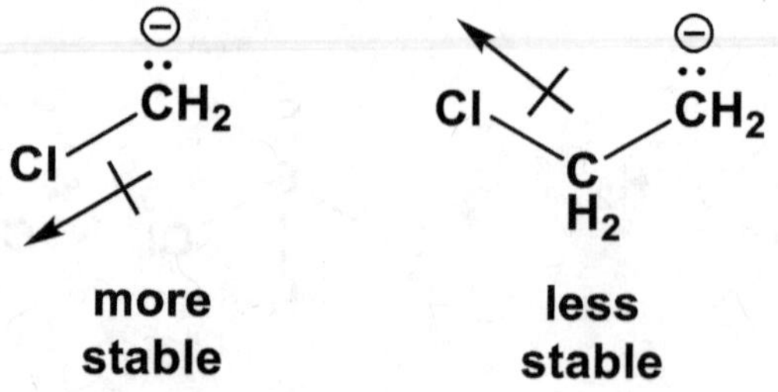


FIGURE 8

Again, we can rank the effects on both electron-rich and electron-deficient carbons (Figure 9).

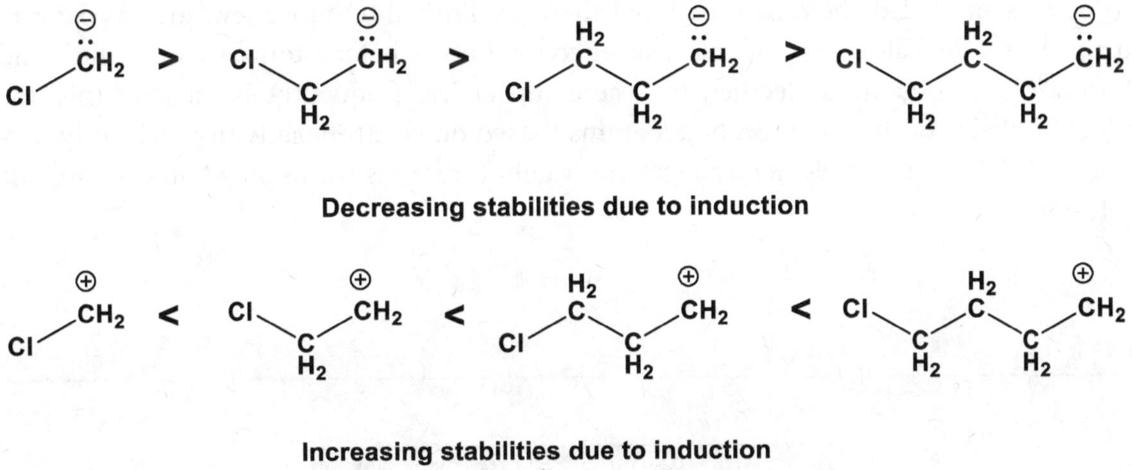


FIGURE 9

At this point, it is important to understand that our rankings of stabilities (and therefore reactivities) are relative and qualitative. As such, we must be careful to always compare apples to apples. That is, we often do not have quantitative values of stability to guide us. For example, in Figure 10 below, we cannot say if stabilization of the excess electron density on carbon is stabilized more by having one electron-withdrawing carbon attached to that carbon or by having two chlorines attached to one carbon away from the carbon with the negative charge. Note that this problem does not exist with resonance, as distance is irrelevant since the electrons "move" throughout the conjugated system.

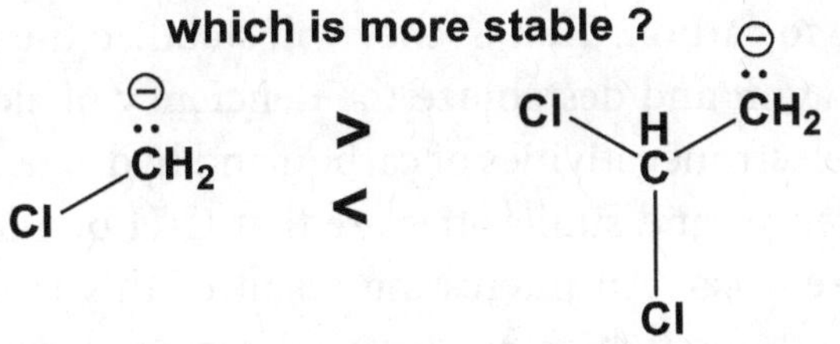


FIGURE 10

Now the good news is that when we consider electronegativities for evaluating inductive effects in organic molecules, we only need to deal with a few atoms. This is because, by definition, organic molecules are made up primarily of carbon and hydrogen. Other atoms can be present (called "heteroatoms"), but these are limited to only a few, such as oxygen, nitrogen, and the halogens. One may also encounter other heteroatoms, such as sulfur and phosphorus, in organic molecules, but these appear less frequently. Because of this, the inductive effects of the common heteroatoms (based on electronegativities) are fairly easy to predict. Table 3 provides electronegativity values for atoms commonly found in organic molecules.

## TABLE 3

| Element | Electronegativity |
|---|---|
| Hydrogen | 2.20 |
| Carbon | 2.55 |
| Nitrogen | 3.04 |
| Oxygen | 3.44 |
| Iodine | 2.66 |
| Bromine | 2.96 |
| Chlorine | 3.16 |
| Fluorine | 3.98 |

Ref: sciencenotes.org

What you will notice is that all of the common heteroatoms found in organic molecules have electronegativity values greater than that of carbon. Therefore, all of these will be electron-withdrawing relative to carbon. That is, they will stabilize the presence of excess electron density on a nearby atom and destabilize the deficiency of electron density. Another important fact is that the electronegativities of carbon and hydrogen are very close in value. The results of these differences and similarities are that C-H bonds are nonpolar and carbon-heteroatom bonds are polar. An interesting result of this is that the family of compounds known as hydrocarbons (which are molecules composed of only carbon-carbon and carbon-hydrogen single bonds) has a very equal distribution of electron densities. And, as is predicted by UDED, they are very stable and non-reactive!

You may then ask if there are any situations in which an induction can provide electron donation. The answer is yes, but only in one special case, which we will describe in the next section on hybridization.

## DISTRIBUTION OF ELECTRON DENSITY VIA HYBRIDIZATION

As stated previously, I consider the effects of hybridization to be a subset of induction. This is because the effects are also best described as affecting "feelings" rather than the actual distribution of electrons, as with resonance. Thus, we can use the same personal experiences given in the description of induction to understand hybridization. That is, the effects are more of a "feeling," and the effects will drop off significantly with distance.

The good news regarding the inductive effect of hybridization in organic molecules is that we only need to consider three hybridization states, which are the most commonly encountered in these molecules. These are *sp*, *sp*$^2$, and *sp*$^3$. In order to understand the effects of hybridization on UDED, we need to first examine these hybrid orbitals more closely.

Let's begin by reviewing how we can determine the hybridization of an atom by inspection of its structure. Table 4 below summarizes the hybridization state of atoms within a molecule based on the number of attached atoms and lone electron pairs.

### TABLE 4

| # ATTACHED ATOMS + LONE ELECTRON PAIRS | PREDICTED HYBRIDIZATION |
|---|---|
| 2 | sp |
| 3 | sp$^2$ |
| 4 | sp$^3$ |

Two important details should always be kept in mind when determining the hybridization of atoms using this formula. The first is that our assignments are based on lone electron *pairs*. That is, lone electrons, as in radicals, do not count. As examples, the hybridization of the carbon in the following molecules exemplifies this difference:

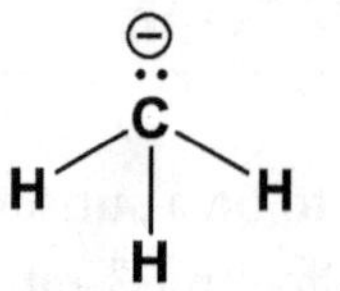

**3 attached atoms +**
**1 lone electron pair = 4**
**hybridization of carbon: $sp^3$**

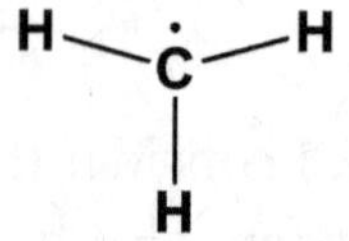

**3 attached atoms +**
**0 lone electron pair = 3**
**hybridization of carbon: $sp^2$**

The other important detail is that this inspection method of counting attached atoms and lone electron pairs allows us to predict the hybridization of an atom, but it is not a guarantee that the atom will hybridize in the way we predict. Indeed, the atoms within a molecule will hybridize in a manner that provides them with the most stability, even if it is different from our prediction. A good example can be realized with amides. In Figure 11, we would assign the hybridization of the nitrogen in the molecule dimethylformamide to be $sp^3$. However, if the nitrogen is hybridized $sp^{3,}$ it will not have any unhybridized *p* orbitals. But, if the nitrogen instead chooses to hybridize as $sp^2$, it will have one unhybridized *p* orbital. Why would nitrogen choose to go against our prediction and hybridize $sp^2$ instead of $sp^3$? The answer, of course, is the significant stabilization that can be achieved via resonance, which distributes the electron density through the conjugated system. Recall that for resonance, the electrons must reside in unhybridized *p* orbitals.

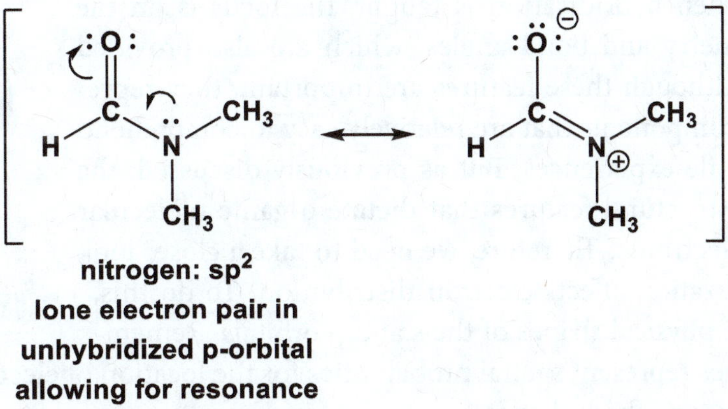


FIGURE 11

Next, let's look closely at the components of the hybrid molecular orbitals. Table 5 summarizes the characteristics of *sp*, *sp*$^{2}$, and *sp*$^{3}$ hybrid orbitals. Of particular importance is the number of unhybridized *p* orbitals present for each hybridized state.

TABLE 5

| HYBRIDIZATION | NUMBER of HYBRID ORBITALS | NUMBER of UNHYBRIDIZED p-ORBITALS | GEOMETRY | BOND ANGLE |
|---|---|---|---|---|
| sp | 2 | 2 | linear | 180° |
| sp$^2$ | 3 | 1 | trigonal planar | 120° |
| sp$^3$ | 4 | 0 | tetrahedral | 109.5° |

Oftentimes, when hybridization is taught, the focus is on the resulting geometry and bond angles, which are also provided in the table. Although these features are important, they represent physical components that are relatively easy to comprehend based on our life experiences. But as previously discussed, the main driving structural features that dictate organic molecular behavior are electronic. Therefore, we need to take a closer look at how hybridization affects electron distribution. To do this, let's review the physical shapes of the s and p orbitals. Remember, these shapes represent spatial probabilities for the location of electrons about an atom. Figure 12 represents these shapes.

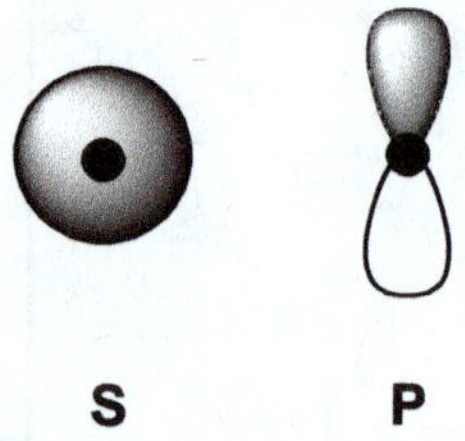


FIGURE 12

In analyzing the location of electrons with respect to the nucleus, we see that electrons in *s* orbitals will be closer to the nucleus than those in *p* orbitals. Of course, the nucleus contains positively-charged protons and uncharged neutrons. This means that the nucleus is strongly electron-withdrawing. Again, this is more of a "feeling" as with induction, since electrons are not moving. Therefore, electrons in *s* orbitals feel a stronger electron-withdrawing effect than electrons in *p* orbitals. For this reason, *s* orbitals are more electron-withdrawing than *p* orbitals.

Now, let's take this reasoning and extend it to hybrid orbitals. We can rank the hybrid orbitals based on their percent *s*-character as shown in Table 6.

TABLE 6

| HYBRIDIZATION | % S-CHARACTER |
|---|---|
| sp | 50% |
| $sp^2$ | 33.3% |
| $sp^3$ | 25% |

Based on our previous reasoning, a hybrid orbital with a higher *s*-character would be more electron-withdrawing than one with less. That is, a *sp*-hybrid orbital is the most electron-withdrawing of the common hybrid orbitals found in organic molecules since it contains the highest degree of *s*-character.

Next, we need to point out something that is not readily explainable based on this concept, but is reality (and therefore must be memorized). The fact is that although *sp* and $sp^2$ orbitals are electron-withdrawing, and *sp* is more electron-withdrawing than $sp^2$, the $sp^3$ orbital is actually electron-donating. This concept is usually explained by the idea of "hyperconjugation," which is not really a conjugative (resonance) effect and therefore can lead to confusion. For the UDED method, it is sufficient to know that an $sp^3$ orbital is electron-donating. In fact, of all the inductive effects realized with organic molecules, it is the only one that presents as electron donating. All others are electron-withdrawing. This is summarized in Table 7.

TABLE 7

| HYBRIDIZATION | UDED EFFECT |
|---|---|
| sp | electron withdrawing |
| $sp^2$ | electron withdrawing |
| $sp^3$ | electron donating |

It is important to point out that the hybridization effects can take place at the atom experiencing an unequal distribution of electron density or due to directly attached atoms. This is akin to the situation with induction. That is, a *sp*-hybridized carbon will greatly stabilize a negative charge on that carbon but will also stabilize a negative charge on an adjacent atom, albeit to a lesser degree (inductive effects decrease significantly with distance).

Let's look at some examples of the effects of hybridization on stability (and therefore reactivity) in organic molecules.

The molecules shown in Figure 13 show excess electron density localized on a carbon. In the first example, the negative charge is on a carbon hybridized *sp*. This hybridization is electron withdrawing; the electrons are held more closely to the nucleus (containing positive charge) and are thereby stabilized more than the second case in which the negative charge is on a carbon hybridized $sp^2$. In the third molecule, the $sp^3$ hybridization of the carbon bearing the excess electron density is electron donating and thereby destabilizing.

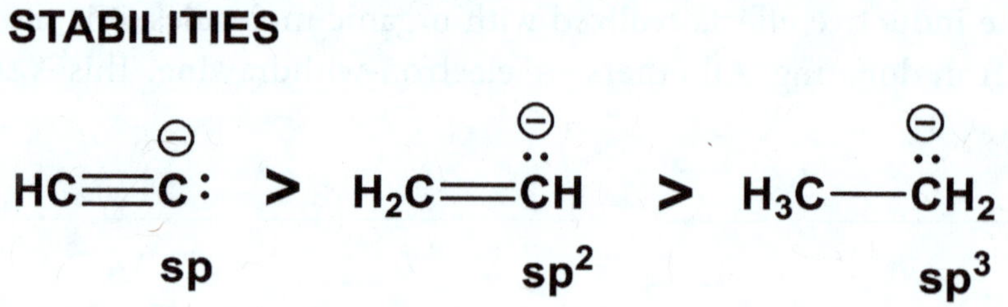


**FIGURE 13**

The molecules shown in Figure 14 show a deficiency of electron density localized on a carbon. In the first example, the positive charge is on a carbon that, by our "rules," is unhybridized. This may not actually be the case, but the atom does have in the second energy level an *s* orbital and three *p* orbitals, two of which are involved in forming pi bonds. In any case, the positive charge on the carbon is very close to the positively charged nucleus, resulting in an unstable configuration. In the second molecule, we assign the hybridization of the electron-deficient carbon to be *sp*, again a hybridization that is electron withdrawing and thereby destabilizing to the electron-deficient carbon. The third molecule represents the most stable of the three. Note that it is not possible to have a full positive charge on an $sp^3$ hybridized carbon since in this case, all orbitals will be filled.

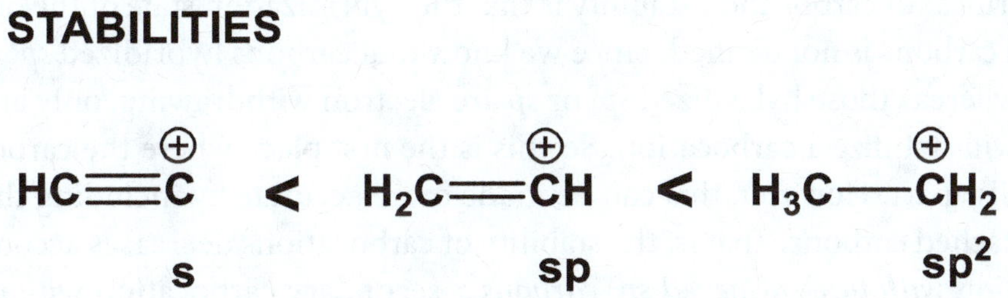


FIGURE 14

Not only is the hybridization of the specific atom which is bearing the unequal electron density important in determining its stability, but also the hybridization of any directly attached atoms. This is important, especially for determining the stability of carbocations. Carbocations are important intermediates involved with unimolecular ($S_N1$ and E1) reactions. Most books will teach that the stability of carbocations decreases in the order of tertiary (a carbon with three attached carbons) > secondary (a carbon with two attached carbons) >> primary (a carbon with one attached carbon), and that primary carbocations will not form under normal conditions due to their instability. This is presented as a "rule" to be memorized and is seldom rationalized other than to offer the previously mentioned concept of "hyperconjugation". The problem with this way of presenting carbocation stabilities is that it will eventually be forgotten (since what we memorize, we will tend to forget). Also, when mindlessly applied, it will lead to inaccurate stability predictions. The good news is that UDED provides a true understanding of this physical reality, which will stay with us for the long term and serves to correct mistakes that often accompany the memorized stability rules.

As with any UDED analysis, we must include an assessment of all three modes of electron distribution: resonance, induction, and hybridization. But for the explanation of the carbocation stability "rule" (tertiary > secondary >> primary), we will assume that there are no resonance or inductive effects to consider. In such a case, the stability effects are due to hybridization. Let's take a closer look at this UDED interpretation of carbocation stabilities.

First, we should examine more closely the definitions of primary, secondary, and tertiary carbon atoms. Very simply, a primary carbon is a carbon with one other carbon attached; a secondary carbon is one bonded to two other carbons, and a tertiary carbon is a carbon with three other carbons attached. Now, the problem with using this terminology in defining the "rules" of carbocation stability is that the hybridization state of the one, two, or three attached carbons is not defined. Since we know that carbons hybridized $sp^3$ are electron donating, whereas those hybridized $sp^2$ or $sp$ are electron withdrawing, only an $sp^3$-hybridized carbon will stabilize a carbocation. So this is the first place where the carbocation stability "rule" falls apart. However, this can be made more accurate by including the hybridization of the attached carbons. That is, the stability of carbocations decreases according to: tertiary carbocation *with three attached* $\mathbf{sp^3}$ *carbons* > secondary carbocation *with two attached* $\mathbf{sp^3}$ *carbons* >> primary carbocation *with one attached* $\mathbf{sp^3}$ *carbon.*

The next problem with the "rule" is that stabilization of the carbocation by other means of electron distribution (resonance and induction) is being ignored. We can justifiably dismiss induction since we know that, relative to carbon, most other atoms found in the majority of organic molecules are electron-withdrawing and therefore would be destabilizing to an atom with a deficiency in electron density. But we cannot justifiably ignore resonance, which, based on the principles of UDED, is the most powerful of the three electron distribution processes. Let's take a look at a couple of examples of how applying UDED will give us a more accurate way to assess carbocation stability. Let's look at the allylic carbocation shown in Figure 15.

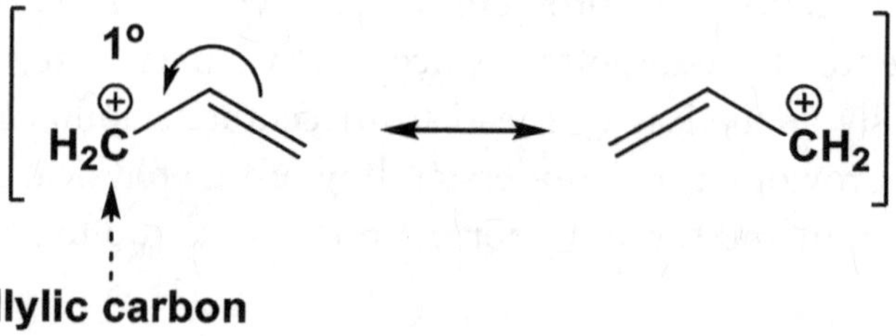


**FIGURE 15**

First, note that the carbocation is primary (only attached to one other carbon). Also, note that the carbon attached to it is $sp^2$, which would be destabilizing (electron withdrawing). In reality, the allylic carbocation is actually *very stable*! This is because of <u>resonance,</u> which serves to distribute the electron deficiency throughout the molecule. And since the principles of UDED state that resonance is much more influential in the distribution of electron density when compared to induction or hybridization, this result (i.e., that *this* primary carbocation is *very* stable) is understood! Even though hybridization of the attached carbon indicates a destabilizing effect, the stabilization due to resonance dominates the overall effect on the carbocation. If we just relied on applying the memorized "rule" of carbocation stability, we would have said that this primary carbocation would never form in the first place!

Another good example is the benzylic carbocation shown in Figure 16.

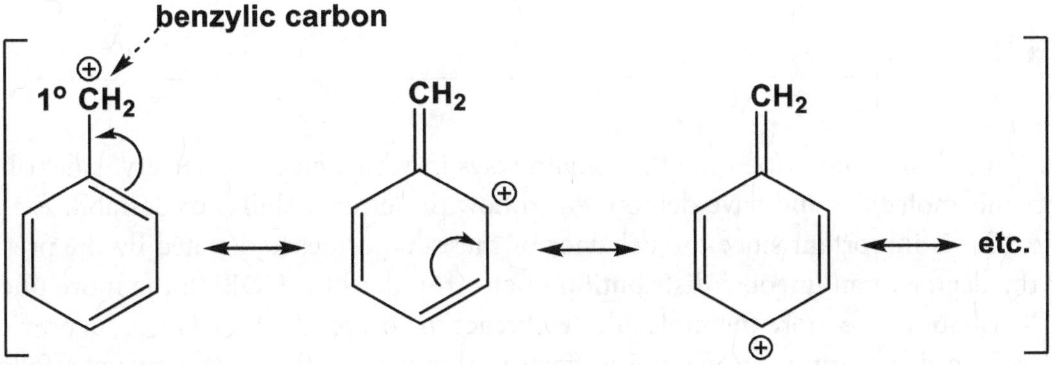


FIGURE 16

Once again, we have a primary carbocation attached to an $sp^2$-hybridized carbon that is very stable due to resonance delocalization into the ring. And once again, the memorized rule that primary carbocations do not form is shown to be incorrect.

Now, should we consider attached atoms that are farther away from the carbon bearing the positive charge when assessing carbocation stability? As was mentioned previously, the hybridization effect is very similar to an inductive effect, and as such, it falls off rapidly with distance. So in these cases, the hybridization of carbons further away from the carbon bearing the positive charge has little effect and is not considered when assessing carbocation stabilities.

A final note on this very important topic of carbocation stability. The actual (real-world) stabilities of carbocations are as shown in Figure 17. Note that a carbon substituted by three $sp^3$-hybridized carbons is more stable than allylic or benzylic primary carbons. Although we cannot predict this exact order of stabilities *a priori* using UDED, we certainly can explain the observed stabilities.

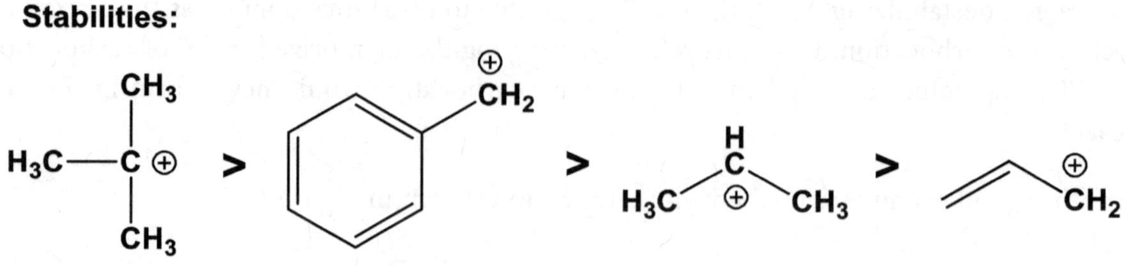


FIGURE 17

In review, we have described the three main ways in which electron density is distributed in organic molecules and have demonstrated how these can stabilize or destabilize a molecule. This is important since the behavior of these molecules is dictated by the presence and the degree of an unequal distribution of electron density (UDED); the more unequal the distribution, the more the molecules will react to correct the imbalance. As previously stated, the order of importance of these effects is: resonance is the most important, followed by induction (including the subset of hybridization; neither being more important than the other). In cases where the resonance effects are opposite to the inductive effects, the resonance effects will predominate.

In the next sections, we will go through a number of examples of applying UDED to understand the behavior of organic molecules. We will use the UDED approach to predict how molecules having an unequal distribution of electron density will react and what products these reactions will produce. In this way, we will be able to comprehend most of the reactions of organic molecules without having to memorize each one (and there are way too many to memorize!).

Although you will certainly encounter many other reactions than those which are covered here during your study of organic chemistry, applying these same UDED principles will lead you to the correct answers 95-98% of the time.

# 3

# APPLICATION OF UDED TO THE ANALYSIS OF ACIDITIES and BASICITIES

It is very likely that you learned about acids and bases in a previous chemistry class. And in this learning, you would have been exposed to the different classifications of said species: Arrhenius, Bronsted-Lowry, and Lewis. These distinctions are derived from the historical development of the understanding of acids and bases. However, we will only focus on describing acids and bases using the Lewis model, which is the most inclusive description. This will hopefully simplify the understanding of these important classes of molecules. What will ultimately become evident is that applying UDED to our understanding of acids and bases will not differ significantly from our understanding of other important organic reactions. As such, 95-98% of the behavior of organic molecules will be understood by applying the same UDED concepts from a singular toolbox. The result will be a unified understanding rather than an endless array of seemingly unconnected chemical behaviors.

First, let's examine acids and bases and classify them as having an excess of electron density or a deficiency. In so doing, we will also label them using more universal terms. That is, terms that can be applied to any molecules, whether they fall under the classification of acids and bases or not. These broader classification terms are nucleophiles and electrophiles.

I find that many students often get these terms mixed up, which leads to confusion in applying the concepts of UDED. Therefore, let's look at how we can know what these terms refer to by deconstructing each name. First of all, both terms contain the suffix "phile". This suffix comes from the Greek word "philos," which means to love or to have an affinity towards or to be attracted to. Some examples are "bibliophile": a lover of books, or "audiophile": someone attracted to high-quality sound reproduction. Moving on to the chemical terms, an electrophile is something that loves electrons. Since electrons are negatively charged, it stands to reason that an electrophile is attracted to excess electron density. Based on our understanding that opposites attract, it stands to reason that an electrophile must exist with a deficiency of electron density. A nucleophile is something that loves the nucleus. So what is there to love in a nucleus? Well, the nucleus contains protons and neutrons. Sadly, the uncharged neutrons only add mass to an atom and, as such, are not all that attractive. However, the protons have a full positive charge. Therefore, a species that is attracted to this positive charge must have excess electron density and is therefore classified as a nucleophile. In summary, electrophiles have a deficiency in electron density and are attracted to nucleophiles, which have an excess of electron density. It is this attraction of these two species, each having an unequal distribution of electron density, and both striving to attain an equal distribution, that dictates 95-98% of molecular behavior.

So applying these categories to the Lewis classification of acids and bases, we can see that Lewis acids can be more broadly characterized as electrophiles (accepting electron density) and Lewis bases as nucleophiles (donating electron density).

With that background, let's look and some organic acids and bases and determine their relative reactivities (stabilities), which of course can be directly related to their pKa's.

Recall that a base is a nucleophile having an excess of electron density and, therefore, a molecule that can donate electrons. The behavior of a base (i.e., its reactivity as a base or its basicity) can be described by the following generic reaction of a base (here shown as reacting with a proton as electrophile):

$$\mathrm{B{:}} + \mathrm{H}^{\oplus} \rightleftharpoons \mathrm{BH}^{\oplus}$$

Note that acid-base reactions are thermodynamically controlled. This means that the reversible reaction will lie to the side (product or reactant) that provides the most overall stability for the system. This feature is important to keep in mind when evaluating basicity with UDED.

Examining this generic representation of the reaction of a base (nucleophile) with an acid (electrophile; in this case, a proton), we can reason the following:

- the MORE stable B: (the base) is;
- the LESS reactive B: is and

- therefore, the more the equilibrium will lie to the LEFT
- and the LESS the base will react with the electrophile (H+)
- meaning that the base (B:) will be LESS basic
- and will therefore have a HIGHER $pK_b$ (or LOWER $pK_a$ for its conjugate acid (BH+)

Before continuing, you must understand these relationships. Therefore, I strongly recommend that you test yourself by filling in the blanks below. If you get any wrong, go back and think about this more until you truly understand.

- the MORE stable B: (the base) is;
- the _______ reactive B: is and
- therefore the more the equilibrium will lie to the _______
- and the ________ the base will react with the electrophile (H+)
- meaning that the base (B:) will be _______ basic
- and will therefore have a _______ $pK_b$ (or ________ $pK_a$ for its conjugate acid (BH+)

With this understanding, let's examine the effects of resonance, induction, and hybridization on the relative strengths of bases. Before we address this challenge, it is important that we move away from just examining one of the electron-distributing effects (resonance, induction, or hybridization) and instead, evaluate ALL of the possible effects to evaluate base strengths. That is, when using UDED to evaluate stabilities (reactivities), we must always evaluate ALL three modes of electron distribution and keep in mind that resonance will always take precedence in describing a molecule's behavior. Also, when evaluating the three means of electron density distribution, the order in which we evaluate these features is inconsequential. The important aspect is that we evaluate ALL three.

Next, we will look at some examples and see if the application of UDED to the analysis of base strengths will provide us with an accurate assessment. In the case of bases (and acids, which will be covered later), we can always check our UDED analysis against reality since experimentally determined $pK_a$ and $pK_b$ values are readily available. Let's begin by analyzing the two bases in Figure 18 using UDED and see if we can predict which would be the stronger (less stable; more reactive) base.

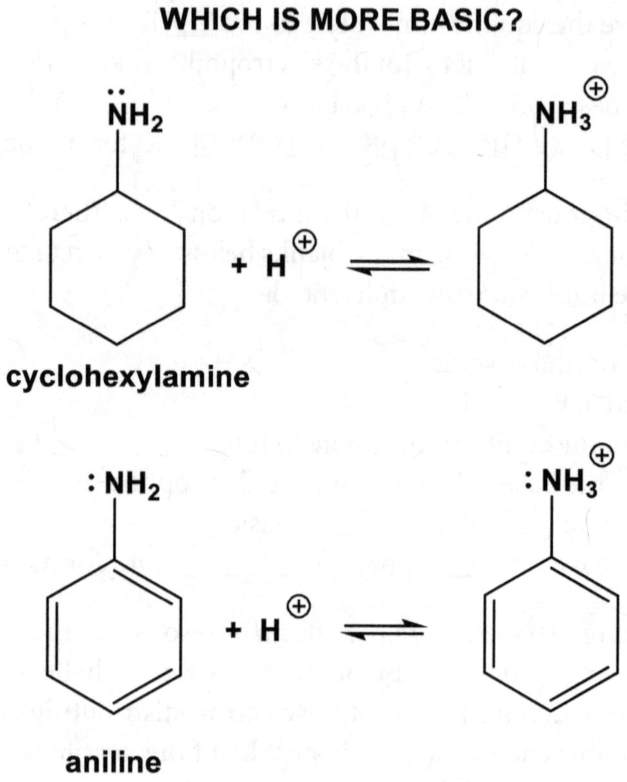


FIGURE 18

We know that a molecule having excess electron density is a nucleophile, and in the case of a base, the molecule will donate that excess in the form of an electron pair to a species that is deficient in electron density (an electrophile). Although the electron-deficient species can be anything, for the examples we will examine, the proton (H+) will always serve as our electrophile.

The first step in this evaluation by UDED will be to identify the nucleophilic center. We define the nucleophilic center as the atom bearing the greatest concentration of electron density. As we will see, this will turn out to be the atom at which reactions occur; the "reactive center". Identifying the center of reactivity is a very important first step in using UDED to predict reactivities. In the case of the molecules in Figure 18, the nucleophilic center would be the nitrogen since it bears a nonbonded (i.e., available to share) pair of electrons. Recall that nonbonded electrons are not critical for the integrity of a molecule (Figure 1).

Next, we need to evaluate the degree to which those electrons are available for sharing. That is, we are looking to determine the relative nucleophilicity (or basicity) at the nucleophilic center. The less distributed those electrons are, the more available they will be for sharing. Also, they will be more reactive since they are more concentrated on one atom (nitrogen) and not shared among other atoms in the molecule. This will translate into a reaction that is driven more to the product side of the reversible base-acid reaction and will result in a stronger base.

Starting with induction, we see that in both molecules, the nucleophilic center is nitrogen. Although nitrogen is electron-withdrawing, the effect of this will be the same in both molecules and will not provide us with a way to distinguish between the two. Therefore, we can eliminate induction as a distinguishing factor in determining their relative basicities.

Next, we examine hybridization beginning with the nitrogen atom (the reactive center). In cyclohexylamine, the nitrogen is hybridized $sp^3$. In this case, hybridization will not stabilize the excess electron density on nitrogen as it is not electron-withdrawing. In aniline, although our "rules" would predict $sp^3$ hybridization for nitrogen, we need to think about this a little more. If the nitrogen in aniline is $sp^3$ hybridized, the lone electron pair would reside in a hybrid ($sp^3$) orbital. As such, it would not be able to participate in resonance (recall orbital overlap for resonance requires unhybridized *p*-orbitals). Thus, in order for the molecule to benefit thermodynamically from resonance, this nitrogen would hybridize $sp^2$, thereby allowing the lone electron pair to reside in an unhybridized p-orbital and thus participate in resonance with the ring. Since $sp^2$ hybridization is electron-withdrawing, the lone electron pair on nitrogen in aniline would be *less* available for donation; that is, it would be less reactive as a nucleophile. As previously discussed, when evaluating the effects of hybridization, we also need to evaluate the effect of groups directly attached to the atom of interest. Looking at the atoms directly attached to the nitrogen, we see that in cyclohexylamine, the carbon attached to the nitrogen is $sp^3$ and therefore electron-donating. This electron donation of the $sp^3$-hybridized carbon would increase the reactivity of the lone electron pair on nitrogen by "pushing" additional electron density toward nitrogen, thereby destabilizing it. In aniline, the atom attached to the nitrogen is an *sp2-hybridized* carbon, which is electron withdrawing and would therefore decrease the electron pair availability for donation. Therefore, our analysis of hybridization says that cyclohexyl amine would be a stronger (less stable; more reactive) base than aniline.

Although we have identified a difference, we cannot stop our evaluation here, but must consider the final way electron density is distributed: resonance. In cyclohexyl amine, there is no resonance delocalization of the lone electron pair available, but in aniline, the electron pair can be distributed throughout the molecule and shared by several atoms (Figure 19). Because of this resonance, the electron pair is much less available for sharing, leading to a reduced nucleophilicity (and basicity) of the nitrogen. Thus, both resonance and hybridization lead us to conclude that cyclohexylamine is a stronger base than aniline. Experimental data show this to be the case with a significant difference in basicities ($pK_b$ values) between the two.

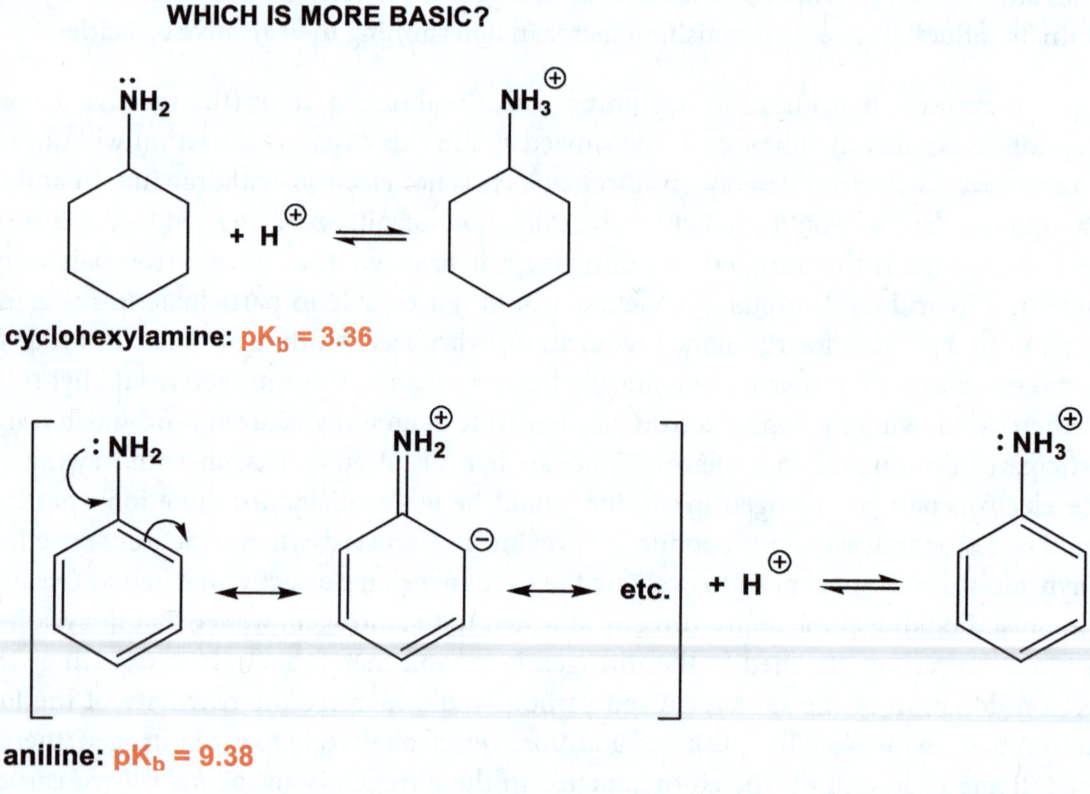


FIGURE 19

Next, I suggest that you practice by evaluating the two bases shown in Figure 20 using UDED and see if you can predict which would be the stronger (less stable; more reactive) base. Once you have done this, read the analysis provided and see if you missed anything.

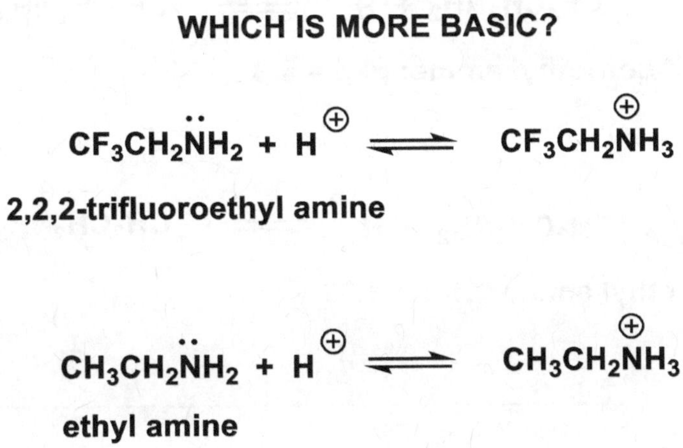


**FIGURE 20**

Once again, we recognize the nitrogen atom as the nucleophilic center. From here we then evaluate the availability of nitrogen's lone electron pair for sharing. As always, we will evaluate each of the three ways in which electron density is distributed. As a reminder, we can do our evaluations in any order as long as we evaluate all three of the ways. Starting with resonance, we see that the lone electron pair in both molecules has no opportunity to be distributed through resonance. With hybridization, both nitrogens are hybridized $sp^3$ and both are attached to $sp^3$ carbons. So neither resonance nor hybridization provides us with a way to distinguish their relative base strengths. This leaves us with induction. Clearly, the molecule bearing three fluorine atoms will be greatly affected by the highly electronegative (electron-withdrawing) effect of the fluorines. This will cause the lone electron pair on nitrogen to be attracted to the fluorines, thereby leaving it less available for sharing with another molecule. Thus, the 2,2,2-trifluoroethylamine will be less nucleophilic (less basic) than ethylamine. This is born out from the experimental $pK_b$ data for these molecules shown in Figure 21.

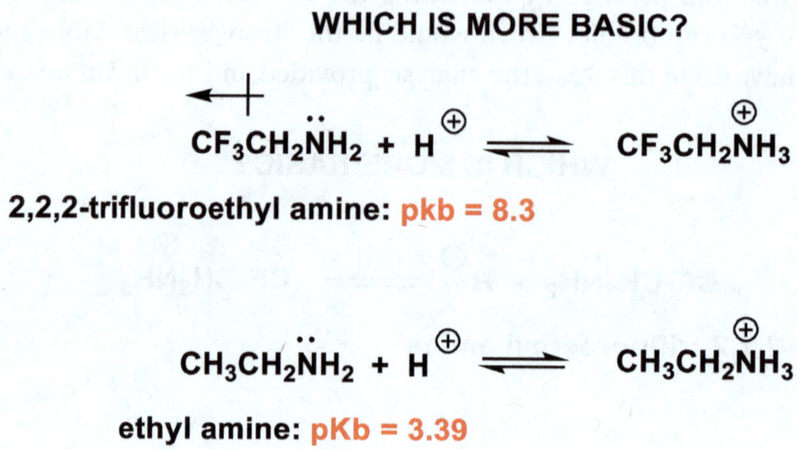


FIGURE 21

Another one to try is shown in Figure 22.

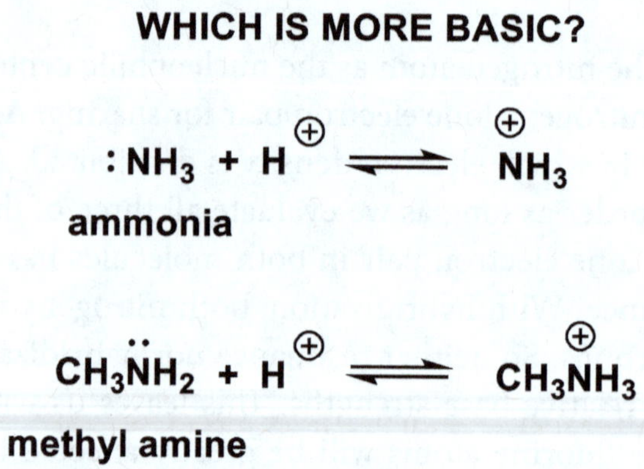


FIGURE 22

As before, we will evaluate the effects of electron distribution to determine relative basicities and will address them in no particular order, being sure to evaluate all three. As with the previous analysis, we can eliminate resonance effects since neither of the nucleophilic centers (nitrogen) has resonance available. Also, there are no strongly electron-withdrawing atoms in the molecules besides the common nitrogen atoms, so we can say that inductive effects do not allow us to distinguish between the two. Now we know that the hybridization of the nitrogens is *sp*³ for both molecules. However, methylamine differs from ammonia by having a *sp*³-hybridized carbon directly attached to the nitrogen. Since we know that *sp*³-hybridized carbons are electron-donating, this effect will place additional burden on the nitrogen bearing the lone electron pair by pushing more electron density towards it, thereby making it more anxious to share its electrons. Therefore, it will be more nucleophilic and behave as a stronger base when compared to ammonia. Therefore, methylamine is predicted to be more basic than ammonia. Again, this is shown to be correct by the data (Figure 23).

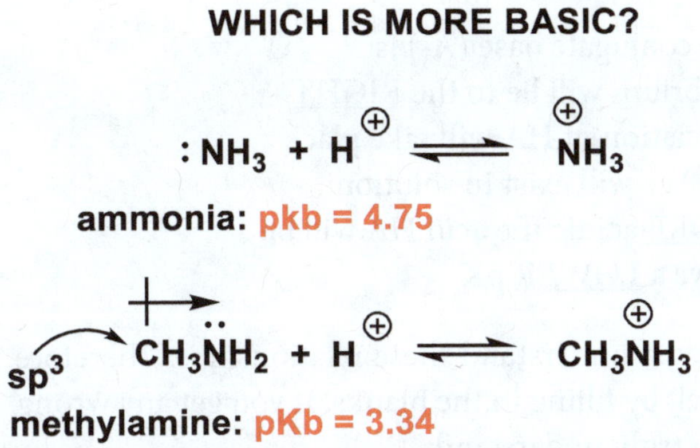


**FIGURE 23**

As a final note, it is important to understand that estimating relative basicities with UDED can be very reflective of reality when $pK_b$ values (or $pK_a$ values of the conjugate acids) are determined in the gas phase. However, when determined in aqueous solution (which is more common), the presence of additional hydrophobic *sp*³-hybridized carbon groups can ultimately inhibit aqueous proton transfer, thereby skewing the predicted relative base strengths. Although we will not discuss this further, realize that UDED may not always be predictive for base strengths determined experimentally in aqueous solution.

We have seen how UDED affects the relative base strength of organic molecules. Next, we will examine organic acids. Although we could evaluate the acidic molecules themselves, I find it more understandable to examine the stability (reactivity) of their conjugate bases. The reason this is a preferable approach is that the conjugate base of an acid has more free electrons. Since with UDED we are evaluating electron densities, the conjugate base provides greater opportunities to evaluate the distribution of electron densities. We can then apply this evaluation of the stability of a conjugate base to the acidity of the acid from which it is derived. Let's explain this reasoning by examining a generic acid dissociation reaction. Recall that acid dissociation is a thermodynamically-driven (i.e., reversible) process:

$$\mathrm{H\text{-}A} \rightleftharpoons \mathrm{H}^{\oplus} + \mathrm{A}^{\ominus}$$

Examining this generic representation of the reversible dissociation of an acid, we can reason the following:

- the MORE stable the conjugate base (A-) is;
- the MORE the equilibrium will lie to the RIGHT
- and the MORE dissociation of HA will take place
- and therefore MORE H+ will exist in solution
- meaning that the MORE acidic the acid HA will be
- and will therefore have a LOWER $pK_a$

Before continuing, you must understand these relationships. Therefore, I strongly recommend that you test yourself by filling in the blanks. If you get any wrong, go back and think about this more until you truly understand.

- the MORE stable the conjugate base (A-) is;
- the ________ the equilibrium will lie to the ________
- and the ________ dissociation of HA will take place
- and therefore ________ H+ will exist in solution
- meaning that the ________ acidic the acid HA will be
- and will therefore have a ________ $pK_a$

With this understanding, let's examine the effects of resonance, induction, and hybridization on the relative strengths of acids. Once again, it is critical that we evaluate ALL three modes of electron distribution and keep in mind the fact that resonance will always take precedence in describing the molecule's behavior. Also, recall that when evaluating the three means of electron density distribution, the order in which we evaluate these features is inconsequential. The important aspect is that each time we evaluate ALL three.

Let's begin by analyzing the two acids in Figure 24 using UDED and see if we can predict which would be the stronger acid.

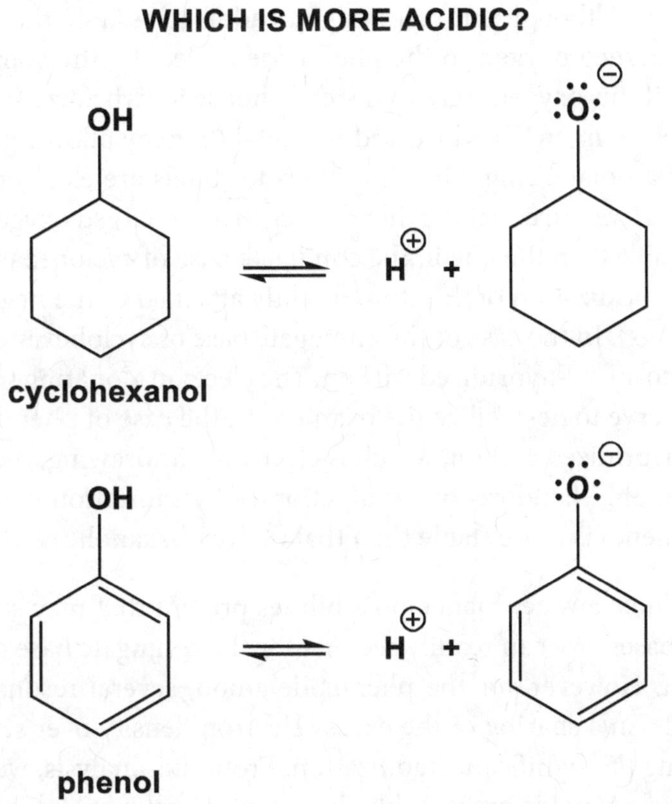


FIGURE 24

The first thing we do is to write the acid dissociation reaction and remind ourselves that this is a thermodynamically driven process. Next, we focus our attention on the conjugate base, rather than the acid itself, and evaluate the stability of this species. Our focus for this evaluation needs to be where the unequal distribution of electron density exists, which is clearly on the oxygen atom, which bears a significant excess as indicated by its formal negative charge. For this evaluation, we will choose to begin by examining induction. In both conjugate bases, the negatively charged oxygen atom has stability due to the electron-withdrawing properties of oxygen (due to its high electronegativity). However, since these induction effects are the same in both conjugate bases, we have no comparative differences.

Next, we can consider hybridization. The oxygen atom in the conjugate base of cyclohexanol is hybridized *sp*$^3$. Although our rules might lead us to assume the same hybridization for the negatively charged oxygen in the phenoxide molecule (the conjugate base of phenol) we realize that if this oxygen were to instead choose to hybridize *sp*$^2$, one electron pair would be able to reside in an unhybridized *p* orbital, thereby allowing it to participate in resonance with the aromatic ring. Since *sp*$^2$ hybrid orbitals are electron-withdrawing and *sp*$^3$ electron-donating, we can conclude that the negatively charged oxygen in the phenoxide molecule is more stable than the one in the conjugate base of cyclohexanol. We can continue to evaluate the hybridization of the atom directly attached to the negatively charged oxygens (AKA: oxanions). In the case of the conjugate base of cyclohexanol, the oxygen atom is directly attached to an *sp*$^3$-hybridized carbon. The electron-donating (or "pushing") effect of this carbon will serve to destabilize the oxanion. In the case of phenoxide, the oxygen is attached to an *sp*$^2$-hybridized carbon, which is electron withdrawing, thereby adding to the stability of the oxanion. Therefore, our evaluation of hybridization can conclude that the conjugate base of phenol is more stable than that of cyclohexanol.

Finally, we will evaluate any resonance possibilities present that may stabilize (or destabilize) the conjugate bases. We can readily see that for the conjugate base of cyclohexanol, no resonance can exist. However, for the phenoxide anion, several resonance structures are available. This results in a sharing of the excess electron density over several atoms within the molecule, leading the significant stabilization. From this analysis, we can conclude that the conjugate base of phenol is more stable than that of cycloxanol due to both hybridization and resonance effects. This means that, based on a UDED analysis, phenol is predicted to be a stronger acid than cyclohexanol. Once again, this conclusion is confirmed by experimental data (Figure 25).

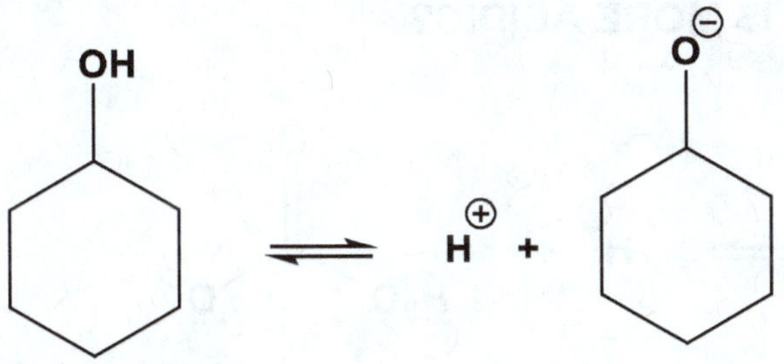

cyclohexanol: pKa = 16

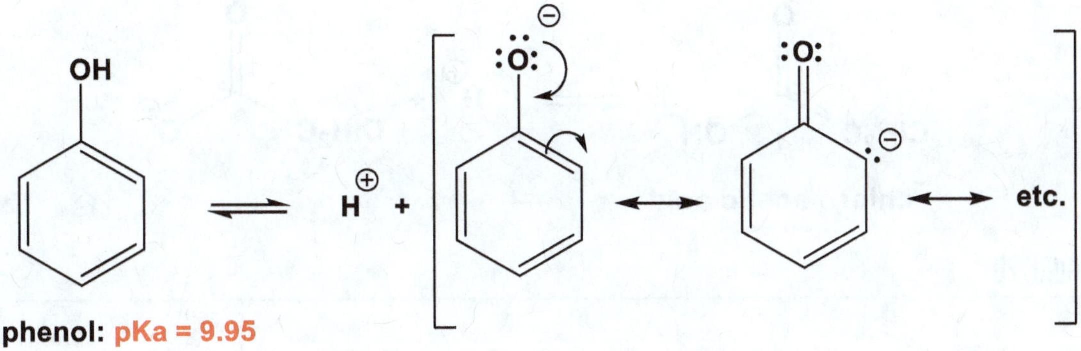

phenol: pKa = 9.95

FIGURE 25

Next, I suggest that you practice by evaluating the two acids shown in Figure 26 using UDED and see if you can predict which would be the stronger (less stable; more reactive). Be sure to first evaluate the stabilities of the conjugate bases and remember that the more stable conjugate base will be formed from the stronger parent acid. Once you have done this, read the analysis provided and see if you missed anything.

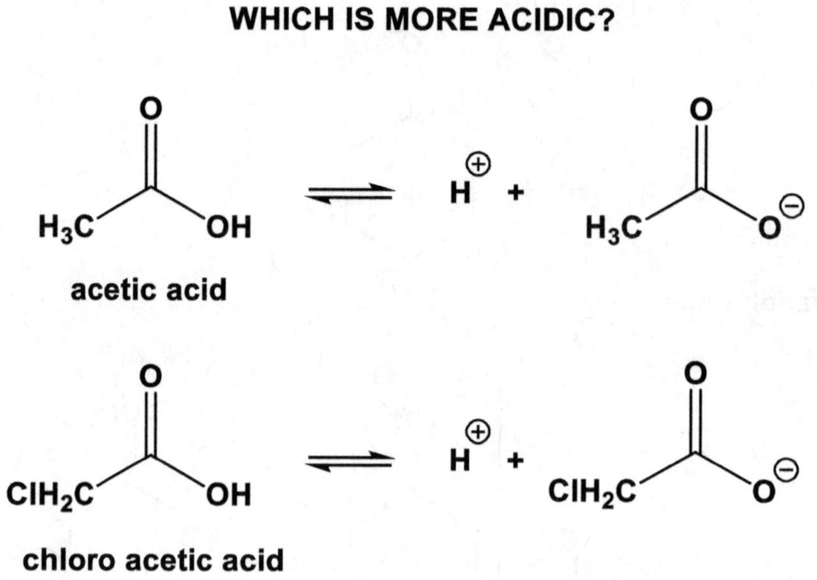


FIGURE 26

Once again, we will evaluate the stability of the conjugate bases and relate this to the acidity of the organic acids. As always, we will evaluate the distribution of the excess electron density of the conjugate bases by resonance, induction, and hybridization.

Beginning with resonance, we see that resonance stabilization is operational in both of the conjugate bases. In fact, the resonance structures are degenerate, making them very good contributors. Figure 27 shows the resonance structures. Notice that I am no longer writing the full Lewis dot structure, since, after much practice, you should be able to realize the existence of the non-bonding electrons. If you are not yet at this point, be sure to go ahead and add the electron pairs.

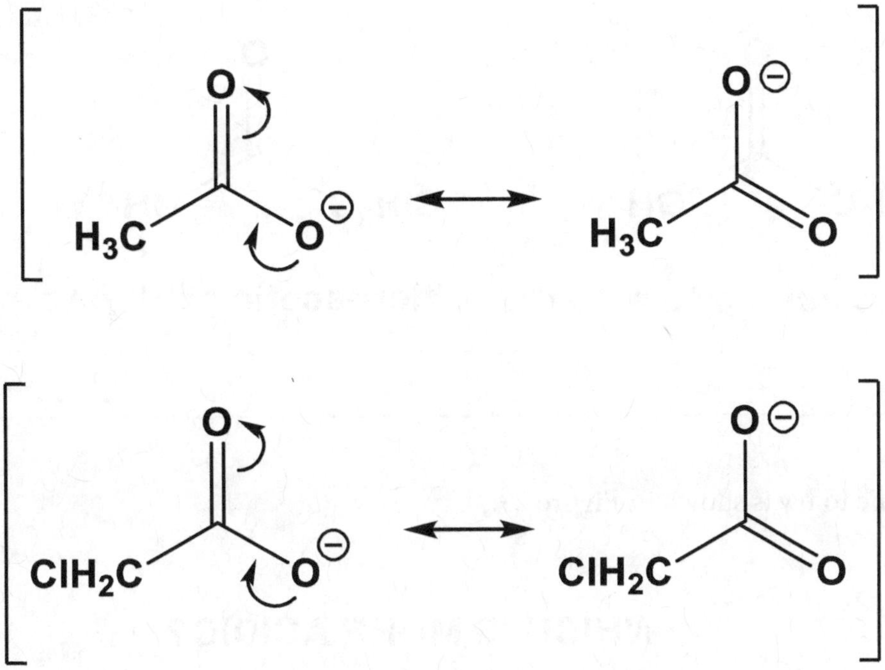


FIGURE 27

Although the resonance stabilization is significant, it is the same for both conjugate bases and thus does not provide an opportunity for us to determine relative stabilities. The same situation is present considering hybridization. That is, both conjugate bases are stabilized by having the negatively-charged oxygen being $sp^2$-hybridized and directly attached to an $sp^2$-hybridized carbon. Thus, we cannot determine relative stabilities of the conjugate bases with resonance or hybridization. This leaves us with induction. The conjugate base of chloroacetic acid gains stability from the electron-withdrawing effect of the chlorine atom. This serves to withdraw excess electron density held by the conjugate base. The conjugate base of acetic acid does not share this electron-withdrawing chlorine atom. Therefore, we can conclude that the conjugate base of chloroacetic acid is more stable than that of acetic acid, and therefore, chloroacetic acid will be more acidic than acetic. Our conclusion is confirmed by experimental data; the $pK_a$ of chlorocacetic acid is lower than that of acetic acid (Figure 28).

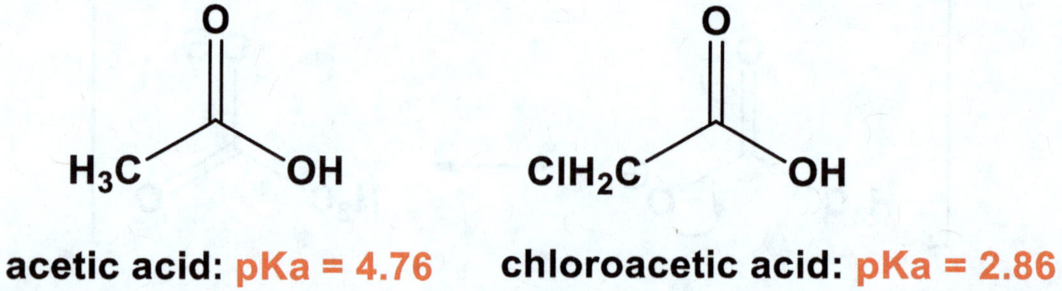


FIGURE 28

Another one to try is shown in Figure 29.

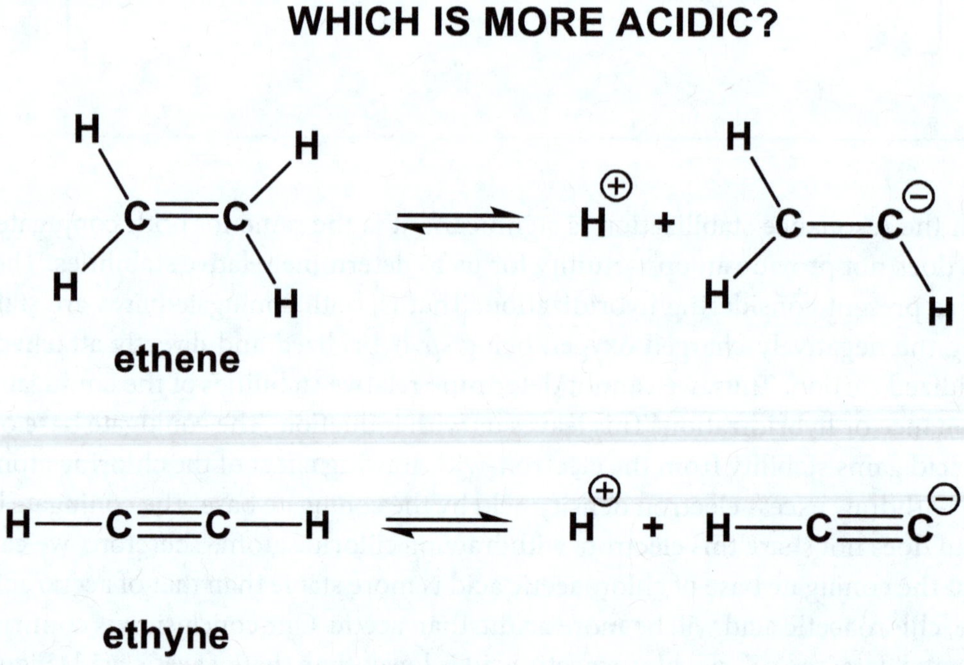


FIGURE 29

When comparing these two conjugate bases, we can see that in both cases, the excess electron density cannot be distributed via resonance (make sure that you can determine this for yourself). Also, the atom bearing the excess electron density (the one with the formal negative charge) in both cases is carbon, and neither has any other electron-withdrawing atoms in the molecule so no difference in electron distribution can be assessed through induction. However, the negative charge in the first example is located on an $sp^2$-hybridized carbon, whereas in the second example, the charge is on an *sp*-hybridized carbon. Since the *sp* hybrid orbital has greater *s*-character, making it closer to the positively charged atomic nucleus, the negative charge will be predicted to be more stabilized. Once again, this is experimentally shown to be correct (Figure 30).

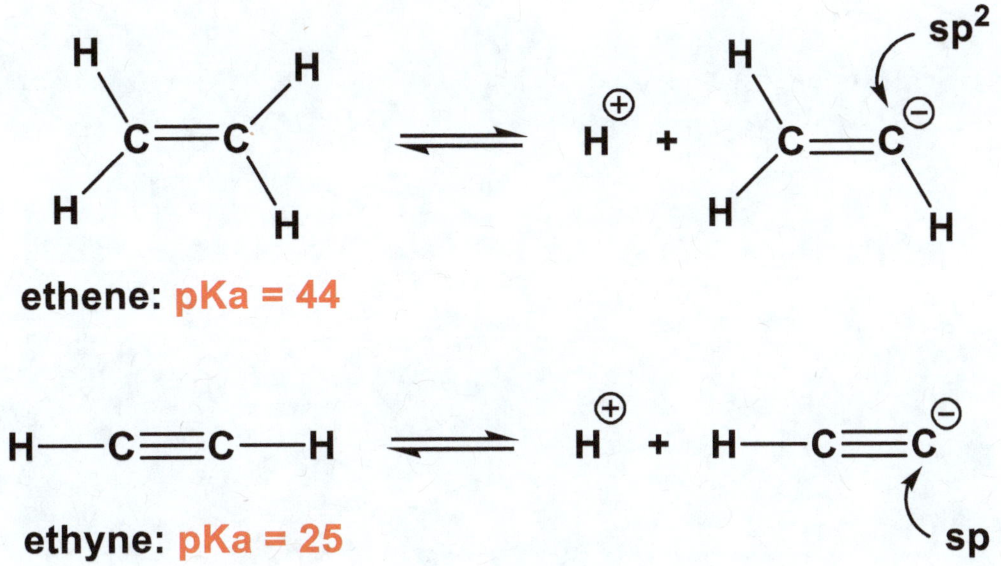


**FIGURE 30**

We can see from these UDED analyses that we can readily assess the stabilities and reactivities of molecules and then apply these assessments to obtain accurate predictions of relative acidities and basicities.

In the next section, we will demonstrate how the use of UDED analyses can be utilized to predict the outcome of other organic reactions. You will see that these analyses are merely an extension of what we did analyzing acid and base strengths. We will use the same UDED toolbox with the same UDED tools. Arguably, this is the most important use of UDED as it will allow for the understanding of reaction mechanisms and the accurate prediction of reaction products without having to rely on brute memorization.

# 4

# APPLICATION OF UDED TO THE ANALYSIS OF ORGANIC REACTIONS

In this section, we will approach the learning of the reactions of organic molecules in a different manner than the way this is usually taught. Instead of describing a reaction in terms of reactants and products and then describing the mechanism, we will first show the reactions and use the concepts of UDED to predict the mechanism and ultimately the products of the reactions.

It is appropriate at this point to remind ourselves of the very basic principle of molecular behavior; that is that opposites attract. We have already discussed this in the introduction by relating this to the behavior of magnets. Hopefully, by now, you should have an intuitive sense of this, but it is important enough to be reminded of this fundamental fact.

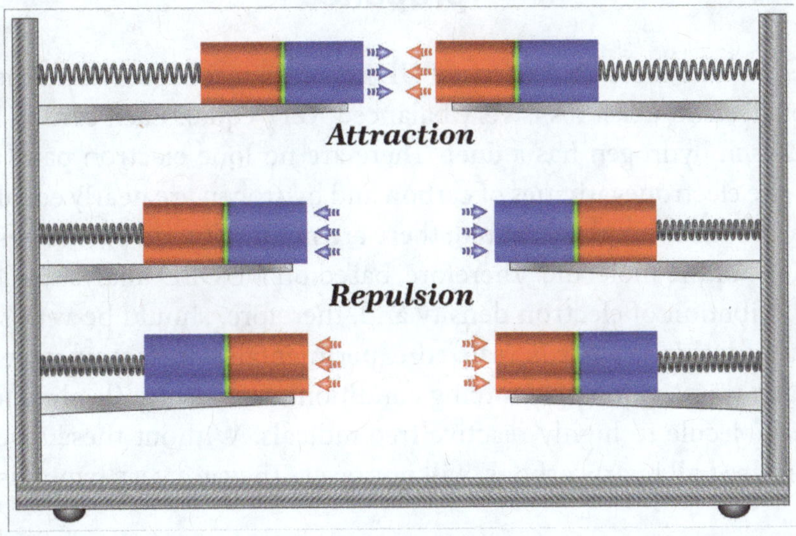


Fouad A. Saad/Shutterstock.com

As we have said earlier, the unequal distribution of electron density causes some molecules to have an excess of electron density (we can compare this to the negative pole of a magnet represented in blue in the above diagram) and other molecules to have a deficiency of electron density (which we can compare to the positive pole of a magnet represented in red). Very simply, molecules will react with one another just as one experiences with magnets; molecules with excess electron density will react with other molecules having a deficiency of electron density. The reason for this is what we referred to earlier as the goal of organic molecules to obtain a utopian society in which all have an equal distribution of electron density.

Before discussing organic reactions, let's take a look at organic molecules that are typically unreactive, except under extreme conditions. The most common of these are the hydrocarbons, which are molecules that contain only singly-bonded carbons and hydrogens. An example is propane:

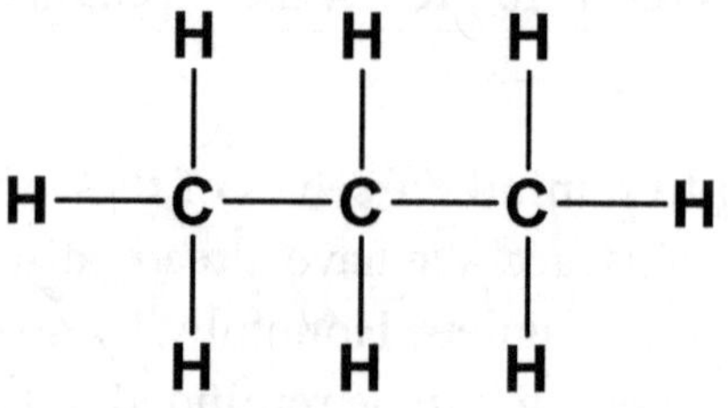


**propane**

If we examine propane through the lens of UDED, we can see that within the molecule, the distribution of electron densities is very balanced, very equal. Each carbon has an octet of electrons, and each hydrogen has a duet. There are no lone electron pairs and no formal charges. Also, the electronegativities of carbon and hydrogen are nearly equal, meaning that the bonding electrons are equally shared; there are no dipoles; no partial positive or partial negative portions of the molecule. Therefore, based on a UDED analysis, this molecule has an EQUAL distribution of electron density and, therefore, should be very stable and very unreactive. Indeed, this is what is realized experimentally. In fact, in order for us to get propane to react, we need to apply forcing conditions such as fire (leading to combustion) or expose the molecule to highly reactive free radicals. Without these severe conditions, propane, and indeed all hydrocarbons, will not react. They are very happy as they are; they are very stable.

However, although the aforementioned reactions of hydrocarbons (combustion and free radical substitution) do exist, they are not readily explained by UDED (except perhaps by considering the UDED of a free radical). For this reason, these reactions fall into the 2-5% of organic chemistry not explained by UDED. Fortunately for those learning organic chemistry, these are minor contributors to the understanding of organic molecular behavior.

Next, we will turn our attention to the majority of molecular reactions, which are readily explicable by applying the UDED method. We'll begin our discussion with an overall analysis of organic reactions based on the UDED concept. Later, we will demonstrate how to apply these analytical steps to the understanding of various reactions.

The general scheme for the UDED analysis of reactions is summarized below:

**UDED leads to the existance of**

**NUCLEOPHILES** + **ELECTROPHILES**

**(molecules that have an excess of electron density)** | **(molecules that have a deficiency of electron density)**

↓

**nucleophiles react with electrophiles via 3 main reaction types:**
**SUBSTITUTION, ELIMINATION, ADDITION**

↓

**PRODUCTS**

The existence of an unequal distribution of electron density leads to molecules that can be classified as nucleophiles or electrophiles. Since we know that molecules will act in an attempt to correct these inequalities, a nucleophile will react with an electrophile in some manner. The three types of reactions that are most common in organic chemistry are described as substitution, elimination, and addition. We will apply the concepts of UDED to understand each of these three types of interactions that will occur between nucleophiles and electrophiles. This approach will allow us to use the universal, straightforward concepts of UDED to understand the myriad of reactions that organic molecules undergo. And rather than trying to memorize the numerous reactions and the products they produce, we can simply apply the UDED method, which will allow us to know them all!

Before examining the three general types of reactions that organic molecules undergo, we will first enumerate the steps that should be used for the evaluation of any chemical interactions. Figure 31 lists these steps. The first step is to evaluate the unequal distribution of electron density within the reacting molecules and identify the nucleophile and the electrophile. Importantly, if our evaluations show that we have two molecules that are both nucleophiles or both electrophiles, we can conclude that *no reaction will take place!* This conclusion is just as important as predicting the products when a reaction will occur.

**THE UDED PROCESS for UNDERSTANDING or PREDICTING CHEMICAL REACTIONS**

**1) Identify nuclepophile and electrophile using UDED**
**2) Identify nucleophilic center (atom bearing most electron density) using UDED**
**3) Identify electrophilic center (atom bearing least electron density) using UDED**
**4) Connect the nucleophilic center to the electrophilic center with a bond**
**5) Have the molecule undergo substitution, elimination, or addition**

**RESULT: 95-98% success!**

**FIGURE 31**

Steps 2 and 3 have us identifying the centers of nucleophilicity and electrophilicity as we did previously for acids and bases. These "centers" are the individual atoms in the molecule where the greatest (for nucleophiles) or least (for electrophiles) electron density exists. As an example, for the ethoxide molecule, the oxygen atom would be the nucleophilic center. Also, for the tert-butyl carbocation, the carbon bearing the formal positive charge would be the electrophilic center:

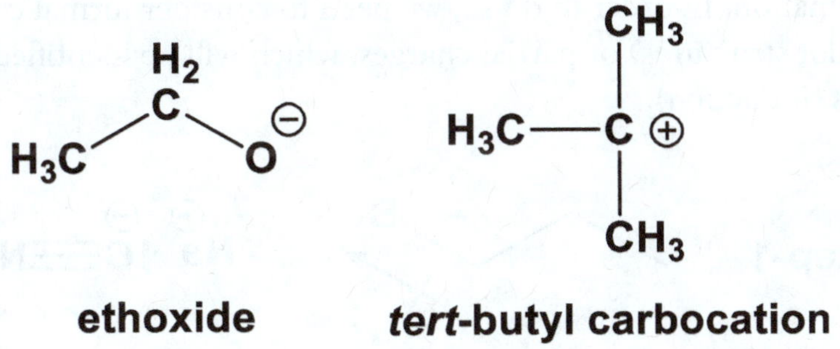


Once the centers of reactivity are identified, we simply connect them with a sigma bond (step 4) and then have the molecule undergo substitution, elimination, or addition. Which of these three main types of chemical reactions will become obvious when considering the structural constraints of the reacting molecules? We will see examples of these as we evaluate each of the reaction types.

With our ability to evaluate UDED molecules and following the steps outlined, we can reliably predict the outcome of the majority (95-98%) of organic reactions. Next, we'll look at examples of each reaction mode.

## SUBSTITUTION REACTIONS

Let's examine the following reaction:

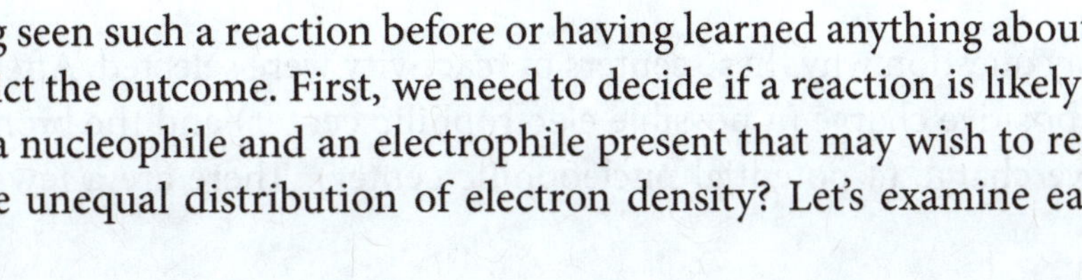


Without having seen such a reaction before or having learned anything about it, we can use UDED to predict the outcome. First, we need to decide if a reaction is likely to occur. That is, do we have a nucleophile and an electrophile present that may wish to react in order to balance out the unequal distribution of electron density? Let's examine each reactant to make this determination. In order to do so, we need to consider formal charges (as determined via Lewis dot structures) or partial charges, which will be identified by considering electronegativities (induction).

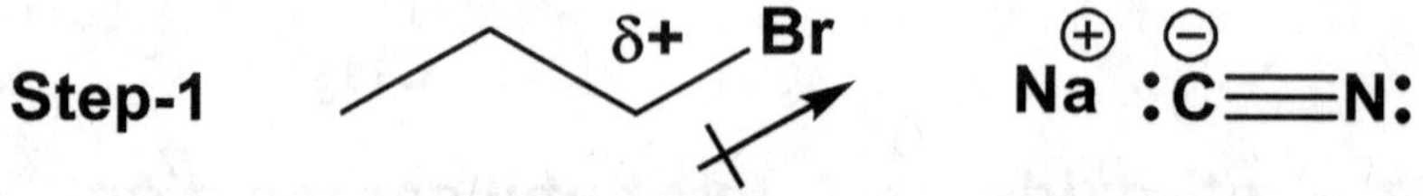


We can see that 1-bromopropane has an unequal distribution of electron density due to the inductive effect of the electron-withdrawing bromine atom. That is, the carbon attached to the bromine has a deficiency of electron density as indicated by both the polarization arrow and the delta plus ($\delta^{+}$) sign. The other reactant, sodium cyanide, has a clear excess of electron density due to the negative charge. Note that the sodium ion is merely a spectator ion, and, as you learned in general chemistry, is very stable with its electron configuration as a +1 ion. This analysis tells us that indeed we have a nucleophile (sodium cyanide) and an electrophile (1-bromopropane) that can possibly react.

For steps 2 and 3, we need to determine the nucleophilic and electrophilic centers. The nucleophilic center is clearly the carbon of cyanide that bears the negative charge. The electrophilic center is carbon-1 of 1-bromopropane, which bears a partial positive charge.

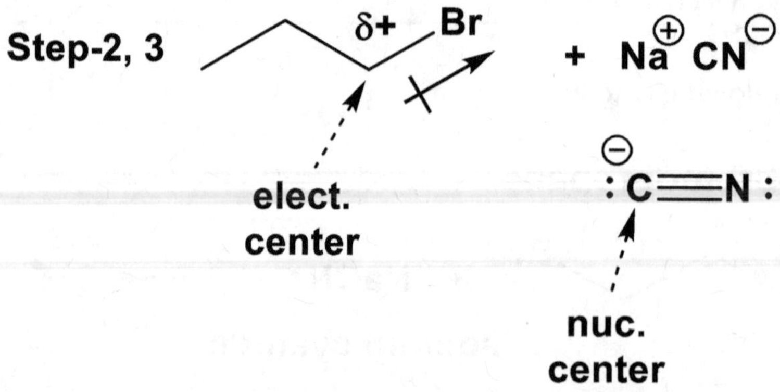


Now, some may question why these centers of reactivity were selected. After all, the sodium ion has a full positive charge (a possible electrophilic center) and the bromine atom has a partial negative charge (a potential nucleophilic center). There are a few reasons for our choices. First of all, these are organic reactions; that is, reactions centered on carbon or hydrogen. Although not always the case, our first instinct should be to consider either carbon or hydrogen as reactive centers. Secondly, as mentioned earlier, we know from our study of general chemistry that sodium +1 is a very stable ion due to its having a noble gas electronic configuration. For this reason, it would not be very likely to obtain extra electrons. Finally, although bromine has a partial negative charge, due to its high electronegativity, it is happy with this situation and would be less likely to react in a way that would require it to share its electron density. Although this rationalization is qualitative, it is the type of thinking that will help in utilizing the UDED method.

In order to complete the UDED analysis of this reaction, we need to address steps 4 and 5. That is, to form a bond between the two reactive centers and have the molecule undergo one of the three main reaction types. Step 4 is straightforward, but what type of reaction should we undergo?

Step-4, 5

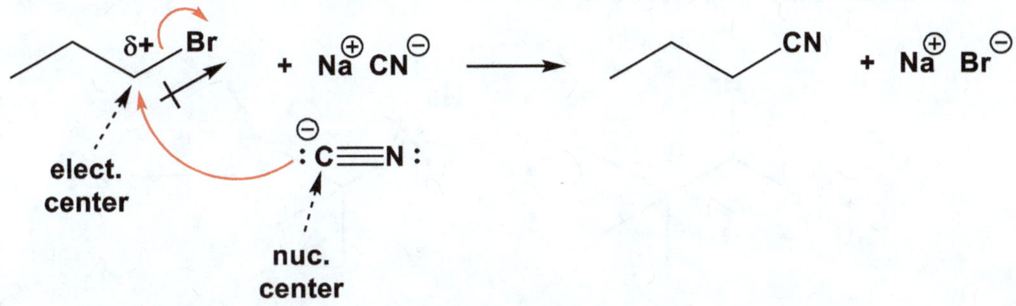


It should be clear that addition is not a viable option. This is because the electrophilic center (carbon) cannot add a fifth bond, as this would violate valence rules. What about elimination? Although possible, this would not be the major reaction pathway. This is something that will be explained as you learn more about these reactions and the difference between good nucleophiles and strong bases, but it is out of the scope of this discussion. Having said this, you will learn that elimination reactions will compete with substitution reactions, which in this case will be the main reaction pathway. This reaction is a classic example of an $S_N2$ reaction type (nucleophilic substitution bimolecular). You will learn about these reaction names in organic lecture classes, but in this discussion, the purpose is for you to learn how to explain the facts and understand the mechanisms using a UDED analysis.

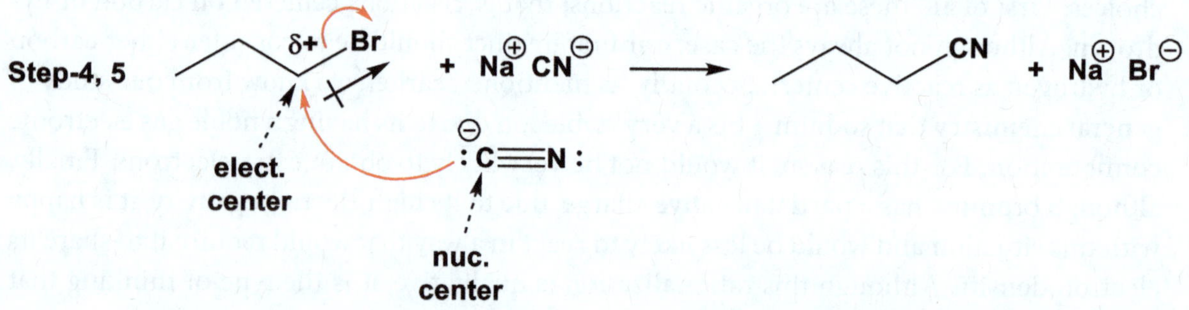


**SUBSTITUTION: CN takes the place of (substitutes for) Br**

## ELIMINATION REACTIONS

Consider the following reaction:

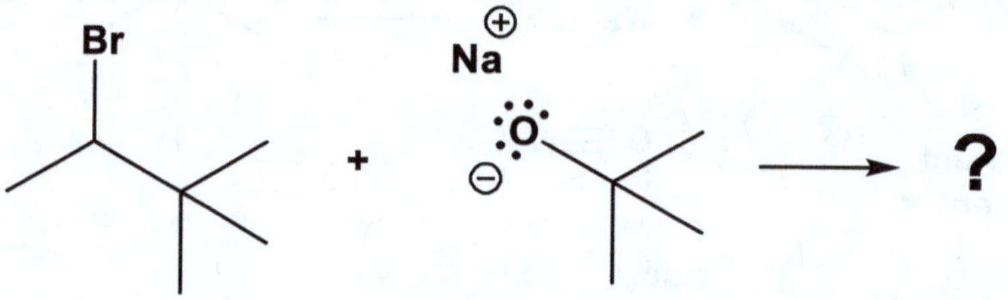


The question is, what will happen when these two molecules are combined? Will anything happen, and if so, what will the outcome be? Again, without having been taught this reaction previously, let's analyze it by applying the UDED method. The first thing we notice is that one of the molecules has an obvious excess of electron density as evidenced by an oxygen atom bearing a full negative charge. Based on our understanding of the non-reactivity of the sodium plus one ion (as previously discussed), we can label this molecule as the nucleophile. This means that in order for a reaction to occur between these two molecules, the other reactant must serve as an electrophile. As in the previous substitution example, we know that the strongly electron-withdrawing behavior of the bromine atom will result in a carbon bearing a partial positive charge. This means that the molecule can indeed serve as an electrophile, and a reaction will occur.

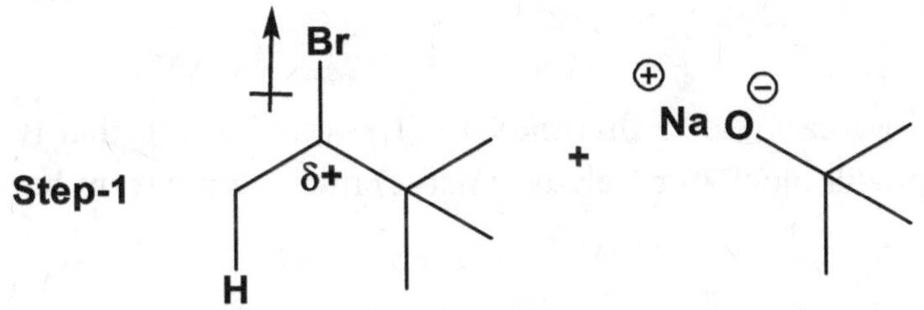


Also, we can identify the reactive centers following steps 2 and 3 of our UDED method.

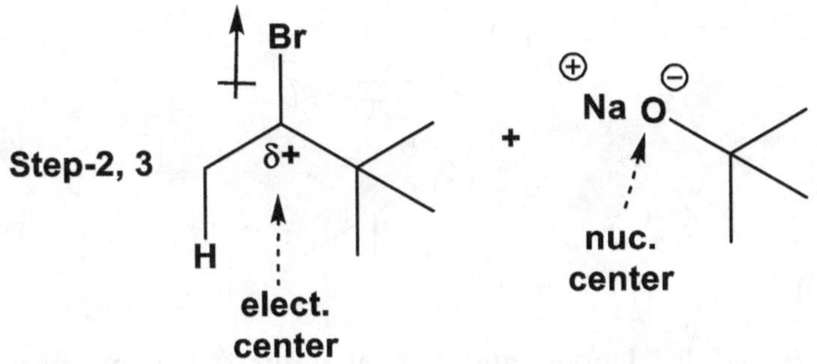


At this point, we could proceed as we did in the previous example and combine the reactive centers to produce a product of substitution. However, there is a problem in this case. The problem is the presence of an adjacent *tert*-butyl (-$C(CH_3)_3$) group which occupies a large amount of available space. That is, it provides significant steric bulk to its surrounding environment.

It is important to pause at this point and discuss steric effects in organic reactions. As was stated in the introduction, sterics, or spatial effects, are relatively straightforward to comprehend. This is because we all have an intuitive feel regarding things that take up space. We know intuitively that we will not be able to fit a large sofa in our compact car when we want to move. We don't need to make measurements; we just know by looking. We can also tell when a container is full that which will prevent us from adding more to it. Translating this intuition to molecules just requires us to work with molecular models, since we cannot see molecules, but most people can translate their spatial intuition to the molecular level with little difficulty. Of course, as pointed out in the introduction, sterics plays a secondary role to electronics in dictating the behavior of organic molecules. This does not mean that we can completely ignore sterics, but rather we need to consider it only after our analysis of UDED.

So in this example, we can realize that the large *tert*-butyl group that is right next to the electrophilic carbon will block our (relatively large) nucleophile from bonding to the electrophilic carbon.

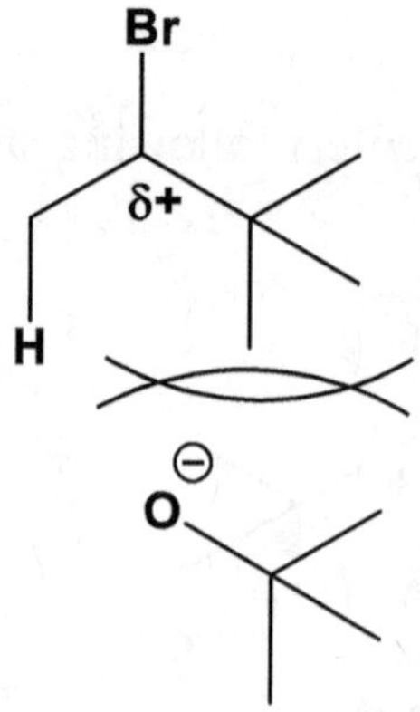


OK, so now what do we do? Understanding that induction is an effect that takes place through sigma bonds, we realize that not only the carbon bearing the bromine will experience a deficiency of electron density, but the atoms attached to that carbon will also experience a reduction in density, albeit to it a lesser degree. We also realize from our spatial perception (and molecular models) that the hydrogen atoms attached to the adjacent carbon are not only small in size but are also situated away from the bulky *tert*-butyl group. We can therefore adjust our selection of the electrophilic center from the carbon bearing bromine to the hydrogens attached to the adjacent carbon.

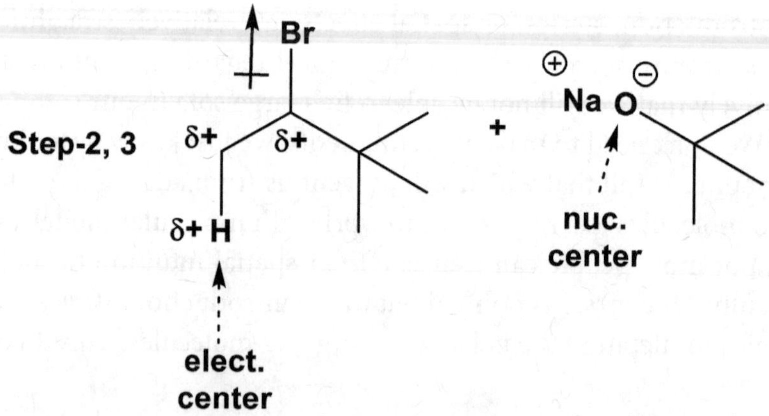

Now we can combine the nucleophilic center with this newly identified electrophilic center. At this point, we have a couple of options regarding the mechanistic pathway for the reaction to take place. We can rule out addition since this would violate the valence rules for hydrogen. What about substitution? Although this is valid, we also know that carbon does not have a high degree of electronegativity. That is, it's not going to be very happy with a full negative charge, which it would need to do if we chose substitution as the reaction pathway (substituting the carbon on hydrogen with oxygen).

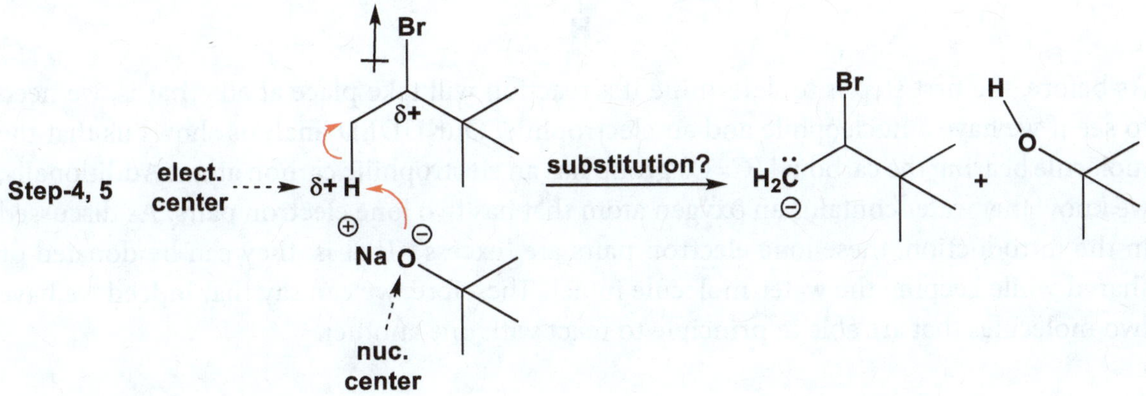


However, we also know that bromine is highly electronegative and would much prefer to hold a negative charge when compared to carbon. Therefore, we can consider a different reaction pathway, namely elimination.

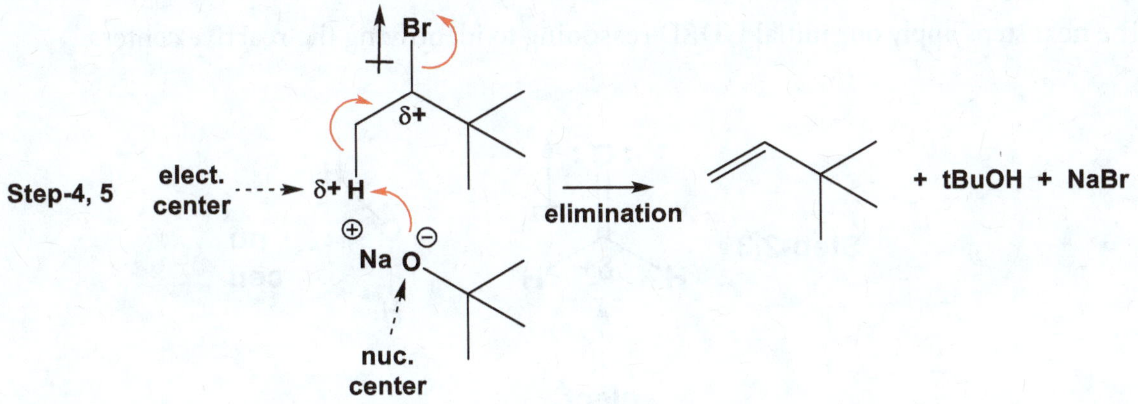

## ADDITION REACTIONS

Consider the following reaction:

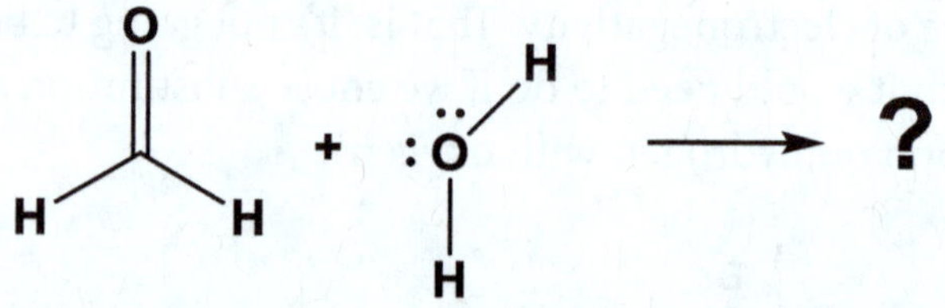


As before, the first step is to determine if a reaction will take place at all. That is, we need to see if we have a nucleophile and an electrophile. Our UDED analysis shows us that the molecule bearing the carbonyl (C=O) group has an electrophilic carbon atom. Additionally, we know that water contains an oxygen atom that has two lone electron pairs. As discussed in the introduction, these lone electron pairs are "excess". That is, they can be donated or shared while keeping the water molecule intact. Therefore, we can say that indeed we have two molecules that are able in principle to react with one another.

Step-1
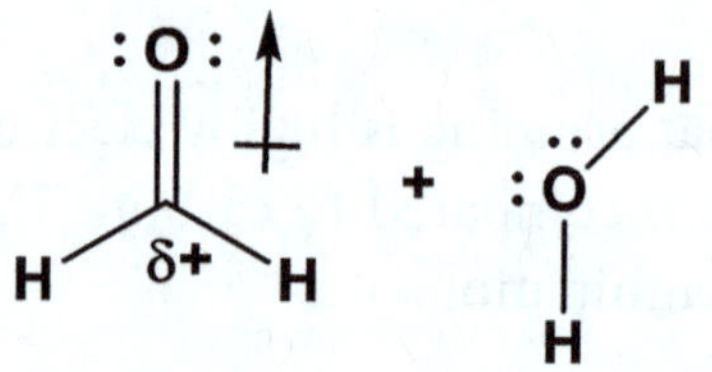


The next steps apply our initial UDED reasoning to identifying the reactive centers.

Step-2,3
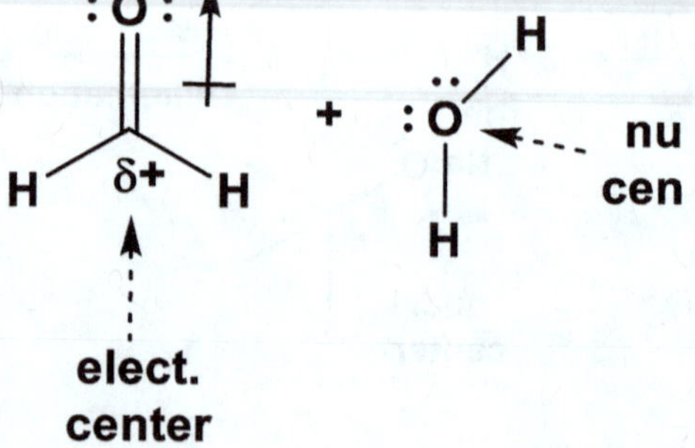

Finally, we connect the nucleophilic center to the electrophilic center with a bond. In this case, since the electrophilic carbon began as $sp^2$ hybridized, it can form another sigma bond to become $sp^3$-hybridized. That is, it can be added to. Therefore, the most likely reaction pathway is addition.

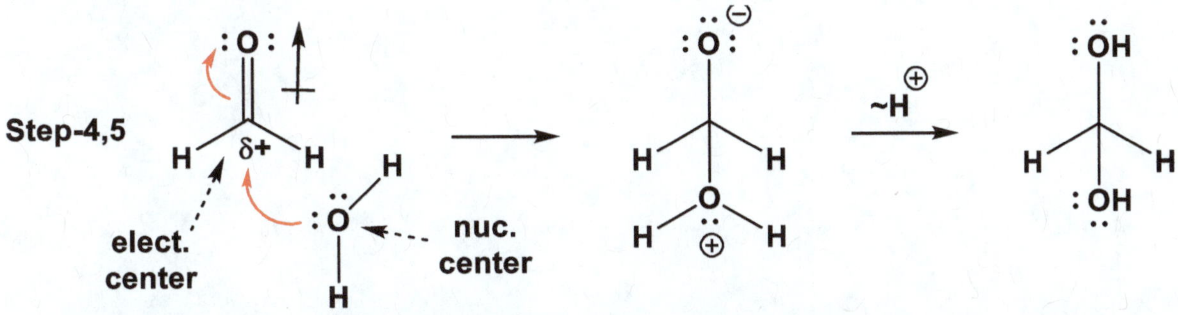


You will note that the final step in the process is to relocate a proton from the positively charged oxygen to the negatively charged one. This "proton transfer" is a very common process and results in a more stable molecule since one of the highly electronegative oxygen atoms no longer has a positive charge. One thing to remember about protons in organic reactions is that "protons are promiscuous". That is, they are not very "loyal" and can readily relocate among atoms, especially electronegative atoms, during the reaction process.

# 5

# CONCLUSION

In conclusion, once you understand the UDED method, be sure to use it anytime you are presented with a new concept or reaction in organic chemistry. If you don't understand something presented in your organic chemistry class, take a look at it through the lens of UDED. 95-98% of the time, UDED will clarify the concept. Be aware that most instructors may not present these concepts from the perspective of the unequal distribution of electron density, but you will almost always be able to understand them with UDED. And understanding is what it is all about!

# 6

# APPENDIX

In this section, we will present some very important topics and tips that will help you succeed in the demystification of organic chemistry. Depending on where you are in your O-Chem journey, these may be completely new to you or will serve as a review. But in any case, it is strongly advised that you become comfortable with these topics.

## REPRESENTATION OF ORGANIC MOLECULES

There are four main ways that are utilized to represent organic molecules. These are Lewis structures, Kekule structures, condensed formulas, and bond-line. Most commonly used are bond line formulas. But it is important to understand that such formulas represent the most abbreviated of the available representations. As such, the reader is required to fill in much of the structure that is left out of the abbreviated drawing. What is most important in the learning process is to always start with the most complete representation of the molecule, and only when you are comfortable, move on to the simpler representations. In these abbreviated drawings, the reader must be able to "see" the presence of atoms and lone electrons even though they are not specifically written. Jumping too quickly to the simpler formulas without being able to visualize the unwritten atoms or electrons is a common cause for confusion and mistakes. Below are summarized the four structural representations.

## Lewis Structures

In Lewis structures, all atoms, electrons, and formal charges are indicated. Being able to construct accurate Lewis structures is key to understanding the behavior of organic molecules via the UDED approach. Unfortunately, the significance of being able to construct Lewis structures is often overlooked, which leads to an early consideration of organic chemistry being "mysterious". The rules for drawing these structures is outlined in every undergraduate organic chemistry textbook. It is strongly suggested that you give this topic extra attention and practice! Again, I recommend the book "Pushing Electrons" by Daniel Weeks (Brooks-Cole publisher), Chapter 1, for additional practice.

## Kekule Structures

With Kekule structures, the Lewis structure is simplified by substituting a line for every two-electron bond. Importantly, every atom and all lone pairs are represented.

## Condensed Formulas

In condensed formulas, the bond lines and lone electrons are left off the drawing, but they are certainly still there! Their presence must be understood by the reader.

## Bond-Line Formulas

These are the simplest molecular depictions in which every apex and terminus of the zig-zag line represents a fully substituted (with hydrogens) carbon. The rest of the structure (lone electrons, carbons, hydrogens) must be imagined by the reader.

Figure 32 shows an example of the various representations of acetic acid.

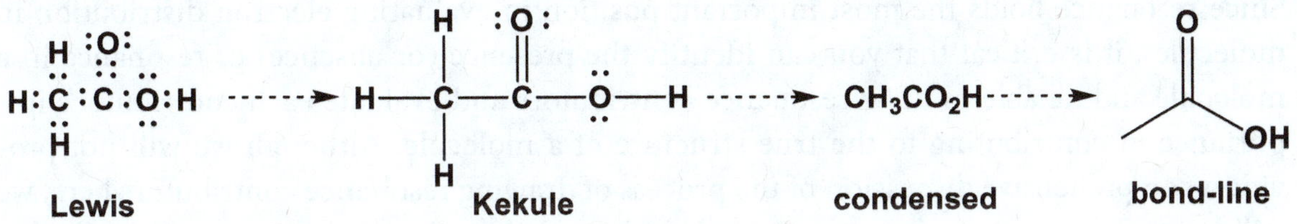


**FIGURE 32**

Note that we are not discussing stereochemistry or the dash-wedge structural representations of stereochemistry. Although stereochemistry is a very important aspect of organic chemistry, it is not something that is part of the UDED method. Stereochemistry will be covered separately in your learning about organic chemistry.

## RESONANCE

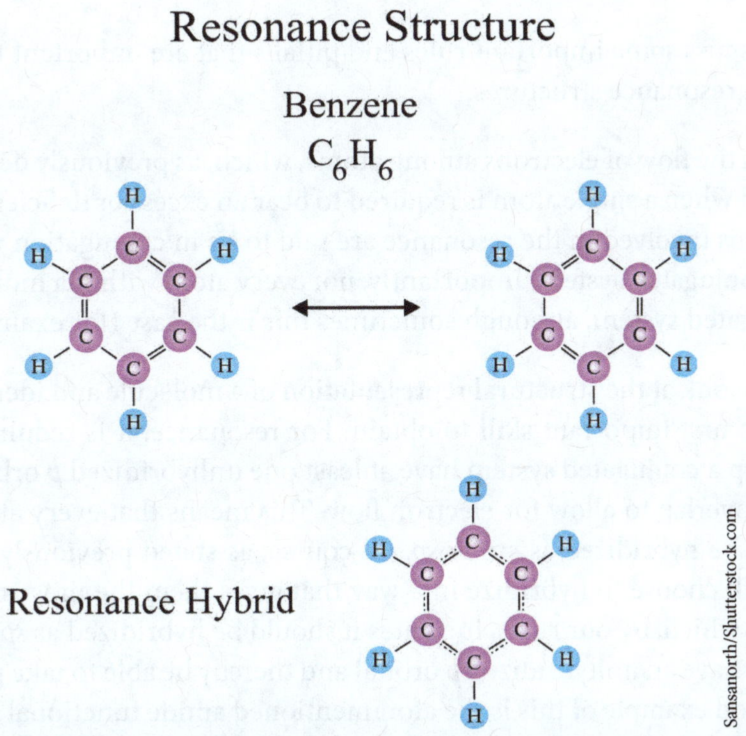

Since resonance holds the most important position in evaluating electron distribution in molecules, it is critical that you can identify the presence (or absence) of resonance in a molecule and be able to draw resonance contributors and evaluate each contributor's importance in contributing to the true structure of a molecule. Although we will not provide a comprehensive discussion of the process of drawing resonance contributors here, we will point out some very important aspects, along with noting some common pitfalls that should be avoided.

First and foremost, it is important to stress that resonance is something that needs to be practiced a lot! One cannot simply read about it and expect you apply it to the UDED method. Think about other life endeavors and how much practice is critical to success. One cannot throw a basketball in the hoop once or twice and expect to be able to play in the NBA. One cannot play a musical scale once or twice and then think they can be a first chair in the New York Philharmonic Orchestra. Hitting a golf ball at the driving range once or twice will not assure you a green jacket in the Masters Tournament.....I think you get my point! And as before, I direct you to the book "Pushing Electrons" by Daniel Weeks (Brooks-Cole publisher), Chapter 2, to get the practice you need. And don't just work the problems once; do them many times each to really get comfortable with resonance.

Next, let's discuss some important rules and pitfalls that are important to be aware of when constructing resonance structures.

Resonance is the flow of electrons among atoms, which, as previously described, relieves the burden faced when a single atom is required to bear an excess or deficiency of electron density. The atoms involved in the resonance are said to be in conjugation with each other and make up a conjugated system. Importantly, not every atom within a molecule has to be part of the conjugated system, although sometimes this is the case (for example, in benzene).

Being able to look at the structural representation of a molecule and identify the conjugated system is the first important skill to obtain. For resonance, it is required that every atom that makes up a conjugated system have at least one unhybridized *p* orbital. This is so the *p* orbitals can overlap to allow for electron flow. This means that every atom of a conjugated system must be hybridized as $sp^2$ or $sp$. Of course, as stated previously, keep in mind that the atoms will choose to hybridize in a way that gives them the most stability. This means that an atom which, by our rules, indicates it should be hybridized as $sp^3$, can choose to hybridize $sp^2$ to have an unhybridized *p* orbital and thereby be able to take part in a conjugated system. A good example of this is the aforementioned amide functional group, in which the nitrogen chooses to hybridize $sp^2$. Finally, all of the atoms of the conjugated system must be consecutive. That is, there cannot be a $sp^3$-hybridized atom in the middle of a conjugated system; electrons can't "jump over" these atoms.

Also of importance in understanding resonance contributors is to realize that not all contributors are of equal importance. In order to determine how much a resonance structure contributes to the true structure, we need to be able to determine the quality of each contributor. Remember that the true structure of a molecule is a hybrid of all of the resonance contributors. However, not all resonance structures will always be equally important. That is, not all resonance structures will contribute equally to the true molecular structure. First, let's review the characteristics of resonance contributors and determine how they affect their quality.

The best resonance contributors will have all atoms bearing an octet of electrons. This is the most important feature to consider when evaluating the contributors. Note that this is even the case for atoms that can expand their octet. A classic example is dimethylsulfoxide (DMSO; Figure 33). In this case, structure **2** contributes more to the hybrid structure of DMSO than structure **1**, even though there exist formal charges on this contributor. This is because all atoms in contributor **2** have an octet of electrons, whereas in **1**, the sulfur atom has ten electrons. Now, since sulfur can expand its octet, structure **1** is a valid resonance contributor, but it contributes less to the true (hybrid) structure of DMSO. This realization shows us that the distribution of electron density in the molecule is better represented by structure **2**.

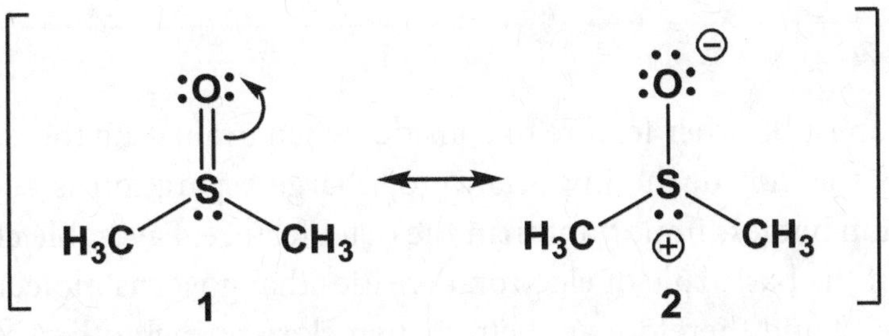


**FIGURE 33**

The next two characteristics that determine the quality of resonance contributors are the formal charge localization and separation. These characteristics are less important than the octet feature and are essentially of equal importance to each other. Charge localization means that a negative charge (high electron density) is best located on an atom with high electronegativity. Let's look at an example (Figure 34). This is the enolate of acetone. All atoms in both structures have an octet, so they are both good in that sense. However, in structure **1**, the negative charge (excess electron density) is on carbon, whereas in structure **2** it resides on an oxygen. Since oxygen is more electronegative than carbon, structure **2** will contribute more to the overall structure of the enolate of acetone. An important note here is that structure **1** is less stable than **2** (due to having the negative charge on carbon rather than oxygen), meaning it is more reactive than **2**. You will see when you study enolate chemistry that although we usually draw enolate structures like **2**, enolates typically will react as nucleophiles in which the nucleophilic center is the carbon, as in structure **1**. This is because this is the more reactive contributor to the overall structure of the enolate.

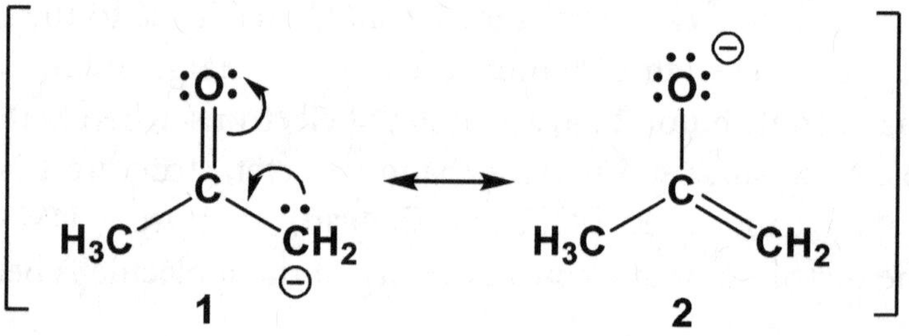


**FIGURE 34**

Charge separation is the other feature to consider when evaluating the quality of resonance contributors. As was noted, the importance of charge separation is equivalent to that of charge localization but less important than the octet feature. Favorable charge separation is a consequence of the basic rule of electron densities that governs molecular behavior. That is, opposites attract and therefore are better when close to each other, and similar charges repel and so are better when separated. In Figure 35, we observe that in structure **1**, the positive and negative formal charges are next to each other. This makes structure **1** a greater contributor to the overall structure of the molecule than structure **2**, in which the opposite charges are separated by two atoms.

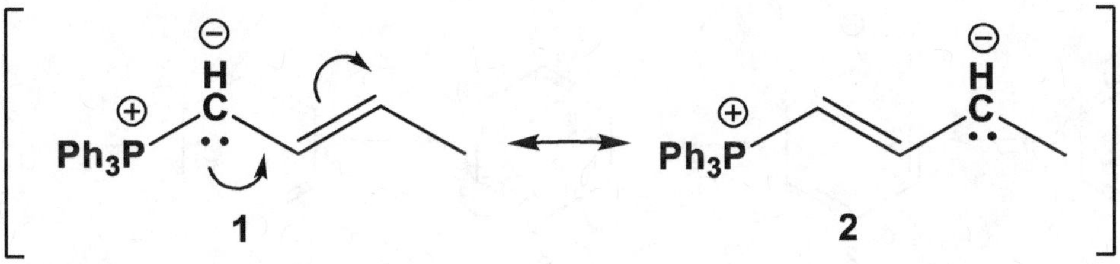


FIGURE 35

Another example of the effect of charge separation on the quality of resonance contributors can be seen in the example in Figure 36. Although both structures have one carbon atom without an octet, they can be distinguished by the degree of separation of the formal positive charges. In structure **1**, the positive charge on carbon is closer to the positive charge on nitrogen than it is in structure **2**. Therefore, structure **2** will contribute more to the hybrid structure of the molecule. Note that in both cases, the positive and negative charges within the nitro group (-$NO_2$) are well-positioned.

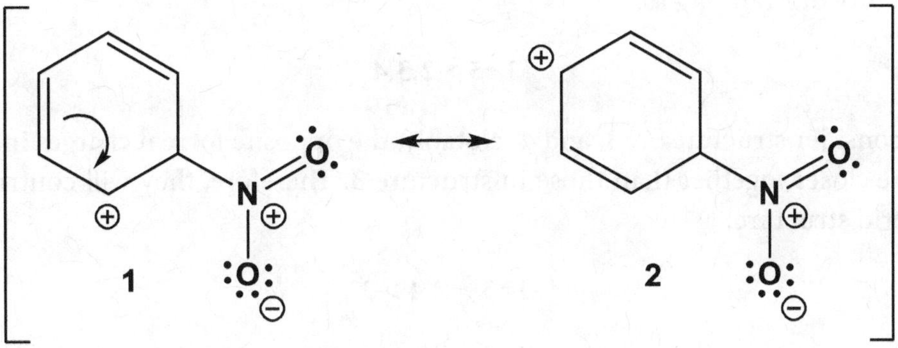


FIGURE 36

Let's evaluate one more example, which will require us to implement a couple of features of resonance contributors. Figure 37 shows several resonance contributors of the molecule benzene (1, 3, 5-cyclohexatriene). Let's rank the five structures in order of decreasing contribution to the overall hybrid structure of benzene.

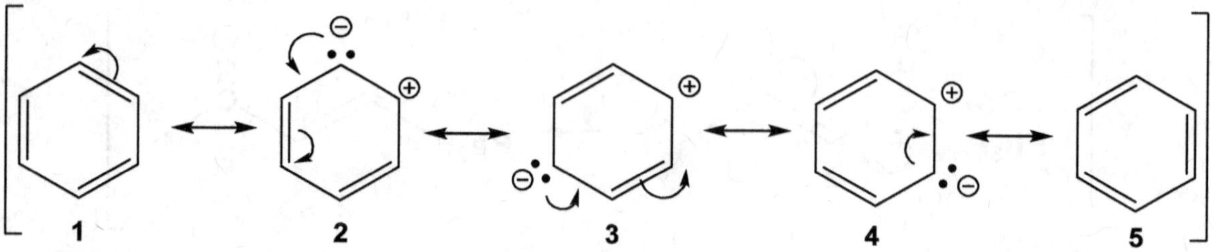

FIGURE 37

First of all, we can see that only structures **1** and **5** have all atoms with complete octets. Therefore, we can start the ranking of the quality of resonance contribution as:

**1,5 > 2, 3, 4**

Now, a new feature of resonance contributors can be noted with structures **1** and **5**. These two structures are identical. In such a case, we term them "degenerate" contributors. And as such, they are considered to be excellent contributors to the overall hybrid structure. So we can modify our ranking as:

**1=5 > 2,3,4**

Next, we consider structures **2**, **3**, and **4**. Notably, the opposite formal charges in structures **2** and **4** are closer together than those in structure **3**. Therefore, they will contribute more to the hybrid structure:

**1=5 > 2,4 > 3**

However, on closer inspection, we see that structures **2** and **4** are degenerate. Therefore, we can finalize our ranking as:

**1=5 > 2=4 > 3**

A few final comments regarding resonance structures are in order. First of all, as previously stated, be sure to practice, practice, practice! Many students, when embarking on this adventure, ask how many resonance structures should I draw for each molecule; when can I stop? The answer is that there can be many valid resonance contributors for a molecule. But when drawing these, learn to include an evaluation of their quality. With experience, you will be able to realize which ones are most important in evaluating UDED effects. As an example, a valid resonance contributor for benzene is shown in Figure 38. Structure **2** is valid because no valence rules have been violated. But it should be easy to see that it would not contribute much to the overall structure since half of the molecule's atoms are lacking an octet. Since the octet feature is not satisfied for this many atoms, it is not a good contributor even though the charges are properly localized. With practice, you will not even bother to draw this resonance structure, even though it is valid.

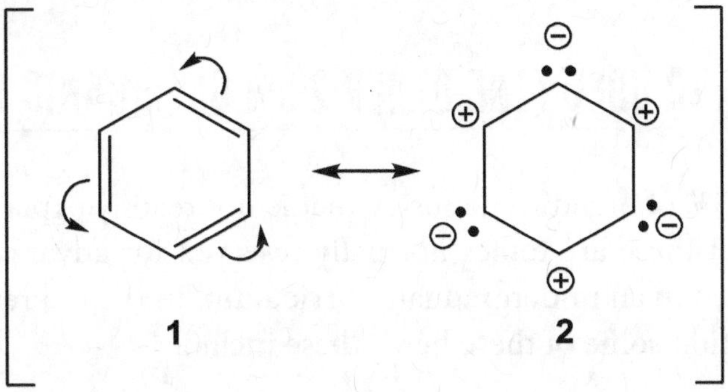

**FIGURE 38**

Another important comment regards the halogens and their involvement with resonance. Although the halogens have several lone electron pairs to participate in resonance, this is often not a consideration. The reasons are twofold. First, the halogens are highly electronegative and will therefore not appreciate sharing their electrons and having to end up with a formal positive charge. More importantly, the overlap of orbitals that is required for resonance only works well when they are of the same energy level. However, aside from fluorine (the most electronegative element), the lone electron pairs on the other halogens are in higher energy levels, causing the overlap, if it were to occur, to be quite poor. For these reasons, we normally do not include halogens as participants in resonance (Figure 39).

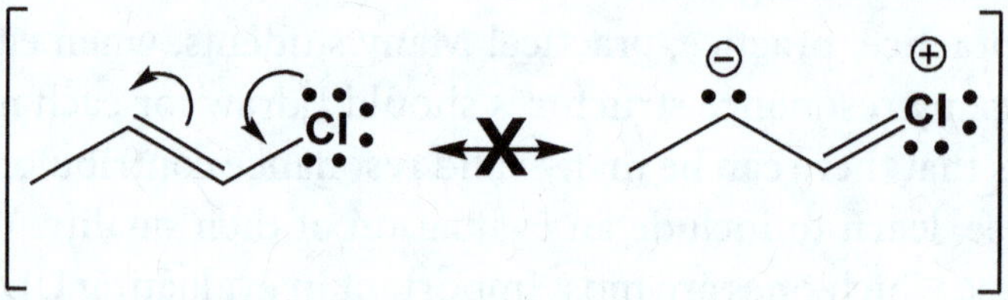


**FIGURE 39**

The final admonition is to be sure to include hydrogens when beginning your practice (that is, use Kekule structures). This will keep you from inadvertently drawing invalid resonance structures with five bonds to carbon, thereby violating valence rules.

## SOME EXCEPTIONS TO UDED (THE OTHER 2-5% OF ORGANIC CHEMISTRY)

So what about the 2-5% of organic chemistry that is not readily explainable using UDED? The good news is that these are topics normally reserved for advanced studies, although some may be addressed in an undergraduate curriculum. In the interest of complete transparency, we will mention some of these here. These include:

- SPECTROSCOPY: Although UDED can be applied to the understanding of spectroscopy (especially IR and $^{1}$H NMR), it is not as useful for predicting $^{13}$C NMR or the IR of carbonyl-containing molecules.
- ELECTROCYCLIC CHEMISTRY: UDED does not apply to the understanding of electrocyclic chemistry (Diels-Alder reactions, Cope rearrangements, etc.).
- PHOTOCHEMICAL and ELECTROCHEMICAL reactions

# 7

# UDED SUMMARY

The following provides a summary review of the UDED method. Once you have read and understood the details of UDED presented in this book, it can be helpful from time to time to review the basic concepts of UDED. You can do this by following the summarized details below.

The distribution of electron density in molecules is mainly determined by three main effects:

## RESONANCE

- Recall that in resonance, electrons are distributed throughout the molecule. The more the electrons are distributed (the more they are shared), the more stable the molecule will be. We "see" this distribution of electron density by drawing resonance contributors.
- Resonance can move electrons toward a site of electron deficiency (electron donating) or from a site of electron excess (electron withdrawing). This dual ability makes resonance a very powerful influence on UDED.
- Since the electrons actually move through the entire conjugated system, there is no distance effect as there exists with induction and hybridization.
- We will most often use bond-line structures in organic chemistry. It is important when drawing resonance contributors using these depictions that you always keep in mind the presence of hydrogens (not written) and the hybridization of the carbons so that you do not violate the valence rules (no more than 8 electrons for carbon).

 ≡ $CH_3CH_2CH_2CH_3$

## INDUCTION

- Remember that inductive effects are simply the "pulling towards" or "pushing away" of electron density (electron withdrawing and electron donating, respectively). Electrons do NOT physically "move" as they do with resonance, which is why the inductive effect has less impact on UDED than does resonance. Nonetheless, this "pulling" and "pushing" does result in UDED. I like to think of induction as more of a "feeling" rather than an action.
- Induction is based on the relative electronegativities of atoms. Since almost all atoms we deal with in organic molecules are MORE electronegative than carbon and hydrogen, almost all inductive effects will be electron-withdrawing. A very important characteristic of induction is that its effect falls off rapidly with distance. This is in contrast to resonance, for which distance is irrelevant to its effect on electron distribution.

## HYBRIDIZATION

- In UDED terms, hybridization simply refers to the location of the electron density relative to the nucleus. Since the nucleus is positively charged, excess electron density will prefer to be closer to the nucleus, and a lack of electron density will prefer to be further away from the positively charged nucleus.
- Since *s* orbitals are closer to the nucleus than *p* orbitals, hybrid orbitals with a greater percentage of *s*-character will be closer to the nucleus. The result of hybridization dictates whether an atom is electron-withdrawing or donating. The following table summarizes these important hybridization details:

| Hybridization of carbon | % s-character of hybrid orbitals | % p-character of hybrid orbitals | Effect on electron distribution |
|---|---|---|---|
| *sp*$^3$ | 25% | 75% | donating |
| *sp*$^2$ | 33% | 66% | withdrawing |
| *sp* | 50% | 50% | withdrawing |

- These effects only apply to carbon atoms, since with other atoms, their electronegativity dominates the effect on electron distribution
- Hybridization can be classified as an inductive effect.
- We know that *sp*$^3$-hybridized carbons are actually electron-donating through an inductive effect known as "hyperconjugation".
- As with induction, the electrons do NOT physically move with hybridization effects.

The relationship between electron distribution, molecular stability, and reactivity is summarized as:

| DISTRIBUTION OF ELECTRON DENSITY | STABILITY | REACTIVITY | SELECTIVITY |
|---|---|---|---|
| **MORE EQUAL** | **MORE** | **LESS** | **MORE** |
| **LESS EQUAL** | **LESS** | **MORE** | **LESS** |

The evaluation of relative acidities and basicities is best accomplished by evaluating UDED and following the basic principles:

- The more stable a base is, the less basic (reactive) it will be.
- For acids, we evaluated the conjugate base. The more stable the conjugate base, the more acidic its parent acid will be.

For Evaluating Reactions:

**UDED** —leads to→ **NUCLEOPHILES and ELECTROPHILES** —which combine via SUBSTITUTION ELIMINATION or ADDITION to give→ **PRODUCTS**

- Molecules react when UDED exists. This is because organic molecules strive for "equality for all," where "equality" means that all atoms have an equal share of electrons....electrons are the "currency" of organic chemistry....a promise of "electrons in every pot*"
- The more equally the electron density is distributed, the more stable the molecules will be (it will be "happier" or more "satisfied"). Remember that the MORE stable the molecule is, the LESS reactive it will be.
- Conversely, the less equal the distribution of electron density, the LESS stable and therefore MORE reactive the molecule is.
- The distribution of electron density is realized in three major ways:
  (1) RESONANCE; (2) INDUCTION; and (3) HYBRIDIZATION.
- The order of importance of these effects on electron distribution is in the order given; i.e., RESONANCE > INDUCTION > HYBRIDIZATION.
- Utilizing the principles of UDED, we can accurately predict the reactivity (or stability) and therefore the behavior of organic molecules.

* This is a reference to the presidential campaign of 1928 in which Herbert Hoover promised that, if elected, there would be a "chicken in every pot and a car in every garage.".....seven months after he took office, the stock market crash of 1929 occurred, which plunged the country into the Great Depression......

# 8

# UDED PRACTICE PROBLEMS

The following questions will test your understanding of the UDED concept and demonstrate how the concept can be applied to the understanding of some of the fundamental principles of organic chemistry. Although the answers are provided at the end of this document, you must give significant thought to each problem before looking at the answers. It is also very important that you understand fully the concepts behind each question.

## PROBLEM #1: LEWIS STRUCTURES AND RESONANCE

Draw Lewis Dot structures for the following molecules:

$$H_3CCO_2CH_3 = H_3C\overset{O}{C}OCH_3$$

$$H_3C\overset{O}{S}CH_3$$

$$H_3CNO_2$$

$$H_3CCOCH_3 = H_3C\overset{O}{C}CH_3$$

$$H_3CCN$$

Which resonance contributor is better (best)? Give the reason for your choice.

Remember the 3 rules for determining the degree of contribution of a resonance structure:

The best resonance contributors have:

(1) all atoms with an octet

(2) formal charges properly localized (negative charges on most electronegative atoms)

(3) formal charges optimally separated (unlike charges close; like charges far apart)

(order of importance = (1) >> (2) = (3)

## PROBLEM #2: LEWIS STRUCTURES AND RESONANCE

Draw the **THREE MAJOR** (best) resonance contributors for $CSO_2^{-2}$ (all heteroatoms attached to carbon). Circle the contributor that would be the best of the three and indicate the reason in no more than 5 words.

## PROBLEM #3: UDED: ASSESSING STABILITIES AND REACTIVITIES VIA RESONANCE

RANK the following molecules according to:

(1) DECREASING stabilities

(2) DECREASING reactivities

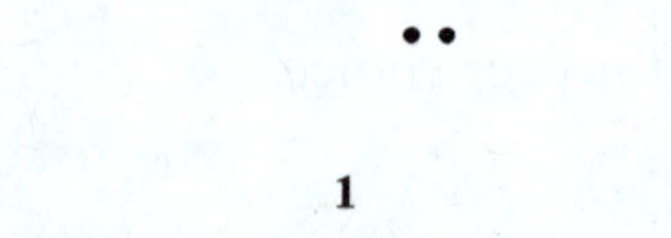

1

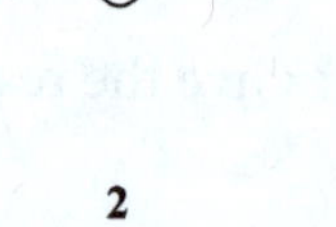

2

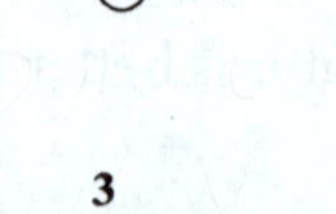

3

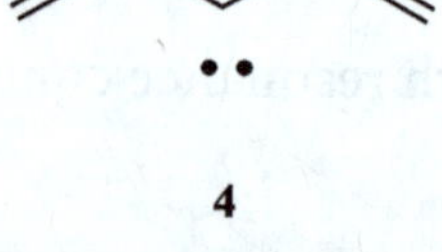

4

Would the molecules be considered nucleophiles or electrophiles?

## PROBLEM #4: UDED: ASSESSING STABILITIES AND REACTIVITIES VIA INDUCTION

RANK the following molecules according to:

(1) DECREASING stabilities

(2) DECREASING reactivities

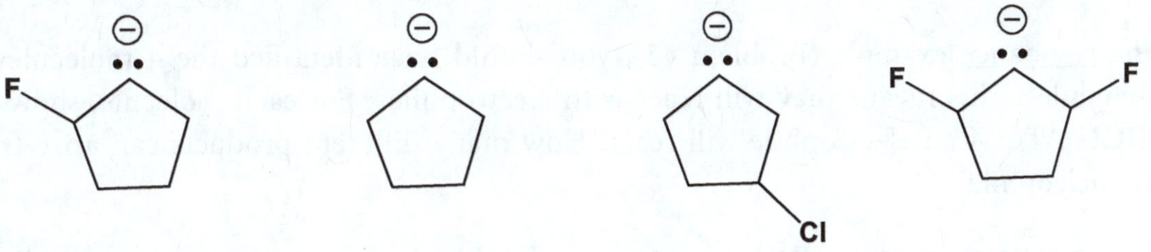


1 2 3 4

Would the molecules be considered as nucleophiles or electrophiles?

## PROBLEM #5: UDED: ASSESSING STABILITIES AND REACTIVITIES VIA HYBRIDIZATION

RANK the following molecules according to:

(1) DECREASING stabilities

(2) DECREASING reactivities

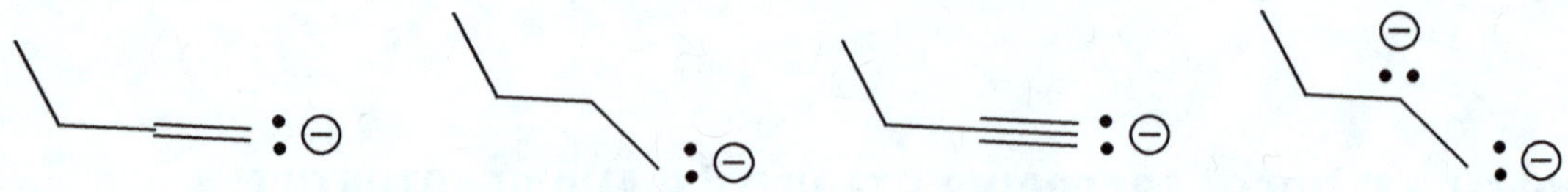

1 2 3 4

Would the molecules be considered nucleophiles or electrophiles?

## PROBLEM #6: REACTIONS OF NUCLEOPHILES

In the <u>resonance</u> example (Problem #3), you should have identified the 4 molecules as nucleophiles. This means they will react with electrophiles. For each molecule, show AT WHICH ATOM an electrophile will react. How many different products can arise from each nucleophile?

In the <u>induction</u> example (Problem #4), you should have identified the 4 molecules as nucleophiles. This means they will react with electrophiles. For each molecule, show AT WHICH ATOM an electrophile will react. How many different products can arise from each nucleophile?

In the <u>hybridization</u> example (Problem #5), you should have identified the 4 molecules as nucleophiles. This means they will react with electrophiles. For each molecule, show AT WHICH ATOM an electrophile will react. How many different products can arise from each nucleophile?

# PROBLEM #7: MORE RANKINGS BASED ON UDED

Use the following choices for answering the ranking questions:

**RANKING OPTIONS (for answer, select LETTER corresponding to correct rank):**

| | | | |
|---|---|---|---|
| A) 1>2>3>4 | G) 2>1>3>4 | M) 3>1>2>4 | S) 4>1>2>3 |
| B) 1>2>4>3 | H) 2>1>4>3 | N) 3>1>4>2 | T) 4>1>3>2 |
| C) 1>3>2>4 | I) 2>3>1>4 | O) 3>2>1>4 | W) 4>2>1>3 |
| D) 1>3>4>2 | J) 2>3>4>1 | P) 3>2>4>1 | X) 4>2>3>1 |
| E) 1>4>2>3 | K) 2>4>1>3 | Q) 3>4>1>2 | Y) 4>3>1>2 |
| F) 1>4>3>2 | L) 2>4>3>1 | R) 3>4>2>1 | Z) 4>3>2>1 |

**Rank the following according to Decreasing Stability and Decreasing Reactivity**

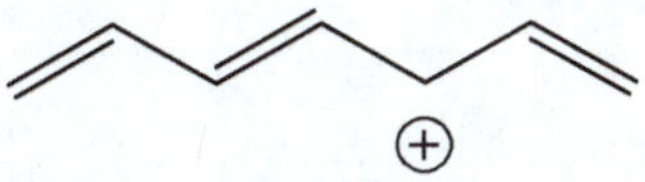

**1**

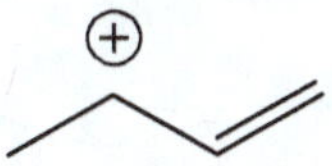

**2**

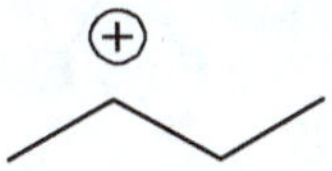

**3**

**4**

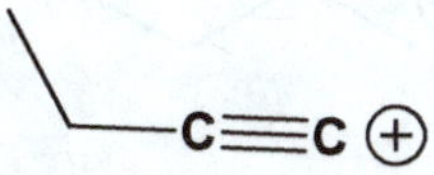


**1**

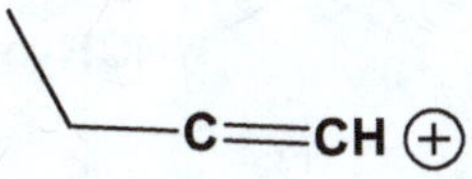


**2**

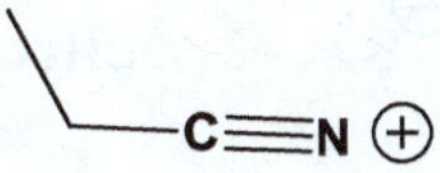


**3**

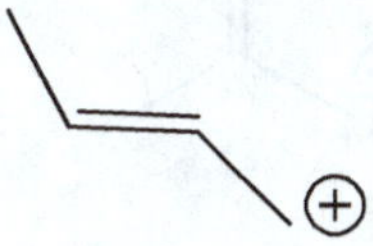

**4**

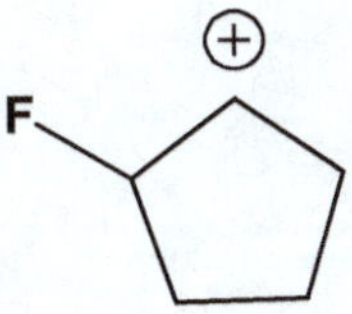


**1**

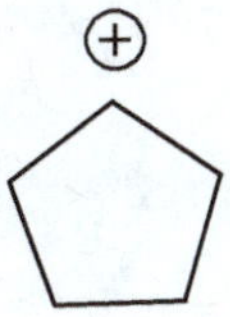

**2**

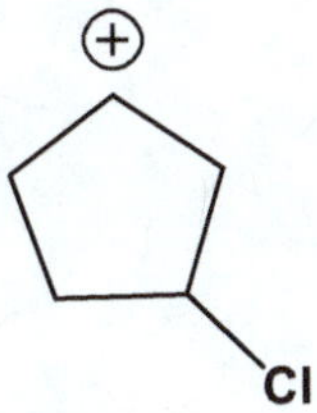


**3**

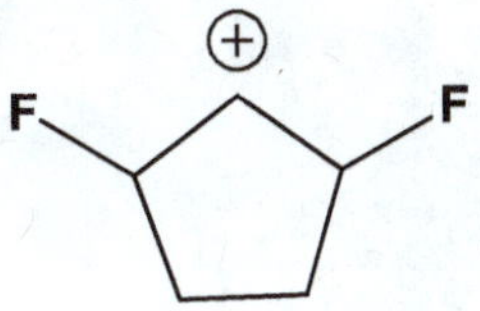


**4**

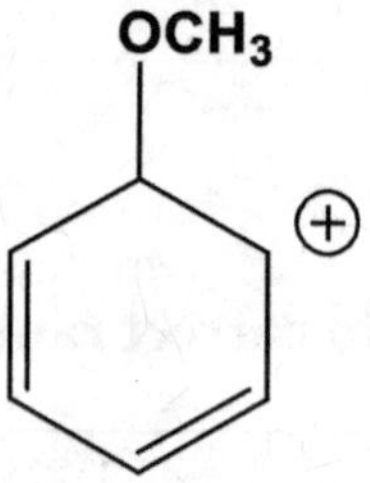


1

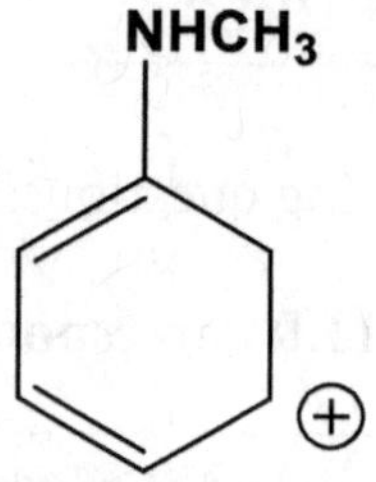


2

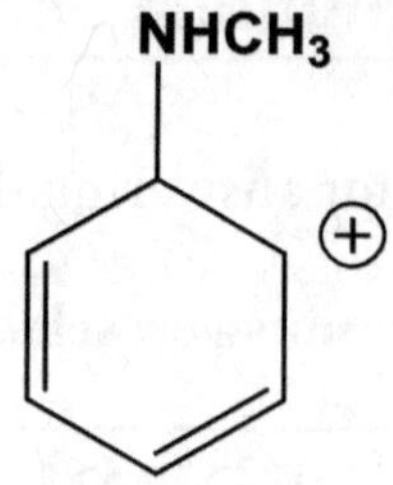


3

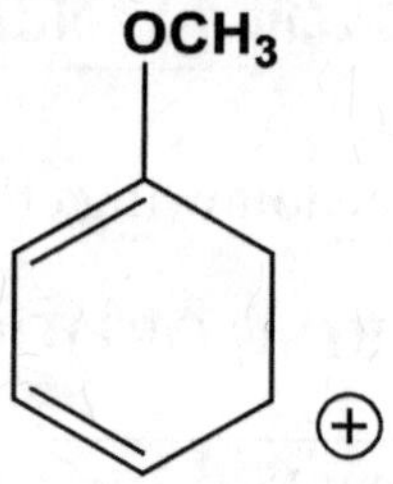


4

**Rank the following in terms of Decreasing Basicity**

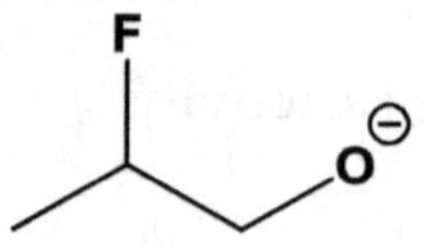


1

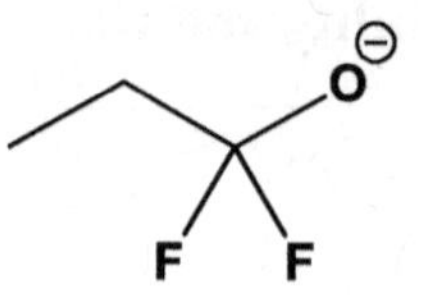


2

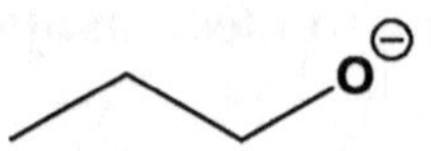


3

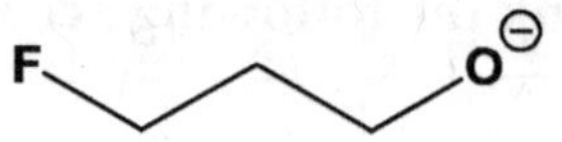


4

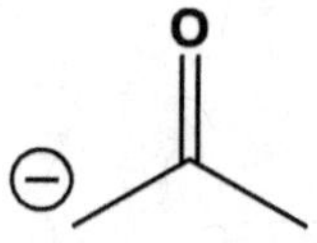


1

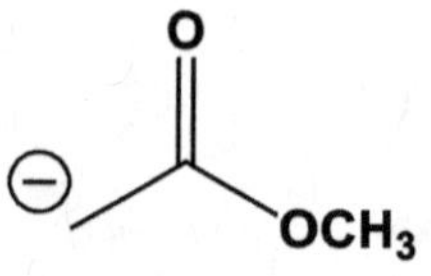


2

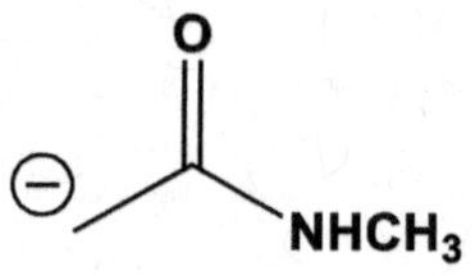


3

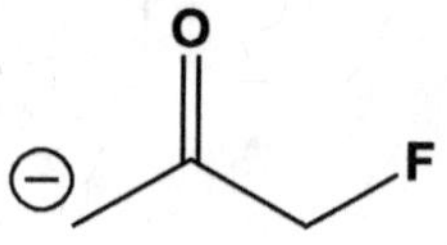


4

**Rank the following in terms of Decreasing Acidity**

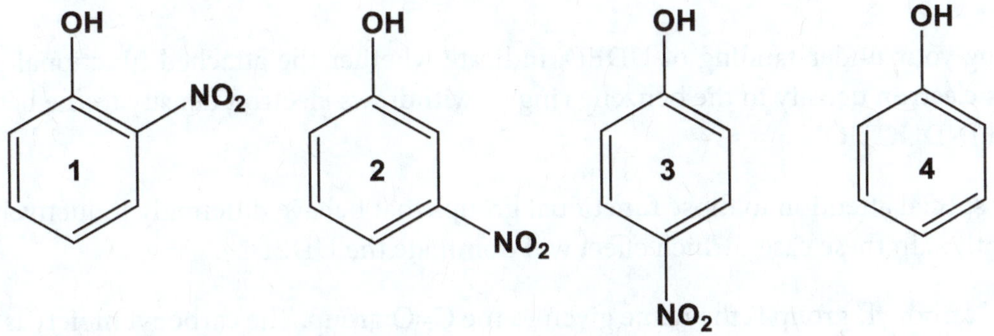


## PROBLEM #8: UDED-ELECTRON DONATING AND WITHDRAWING

You should be familiar with the benzene ring (shown below). This is a very important molecule in organic chemistry and represents the "aromatic" functional group. You will learn much more about this important functional group in 140B but in 140A you will see it present in many molecules.

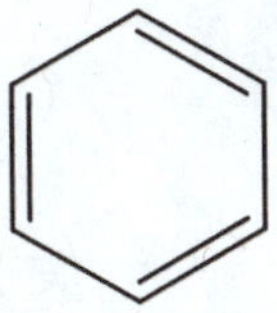

BENZENE

For the following substituted benzene rings, answer the following questions:

(a) Name the functional group attached to the benzene ring (not a part of UDED but good practice!).

(b) Using your understanding of UDED, indicate whether the attached functional group donates electron density to the benzene ring or withdraws electron density to the benzene ring via RESONANCE.

(c) Using your understanding of UDED, indicate whether the attached functional group donates electron density to the benzene ring or withdraws electron density to the benzene ring via INDUCTION.

(d) Pay special attention to those functional groups that behave differently inductively and resonantly.....in these cases, which effect will dominate the UDED?

(e) The "carbonyl" group is the name given to the C=O group. The carbonyl moiety is not a functional group by itself but is an integral part of many functional groups. Identify all the functional groups that contain a carbonyl moiety.

(f) From your work in question #6, is it correct to say that the ester and amide functional groups always withdraw electron density from the benzene ring?

(g) RANK the following in order of decreasing electron density in the attached benzene ring:

8; 14; 17; 20

8; 15; 17; 19

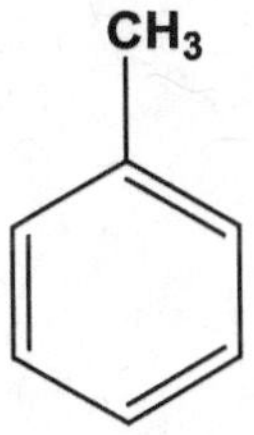


1

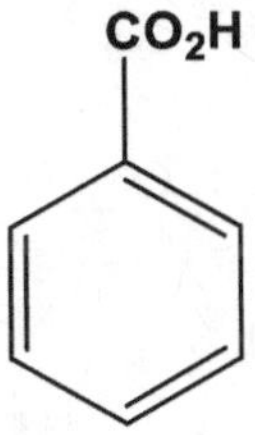


2

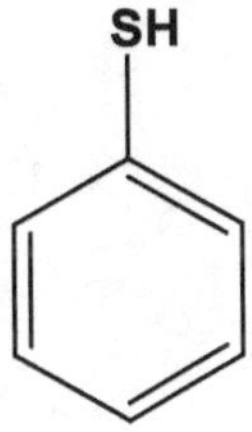


3

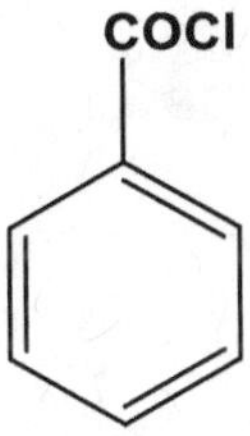


4

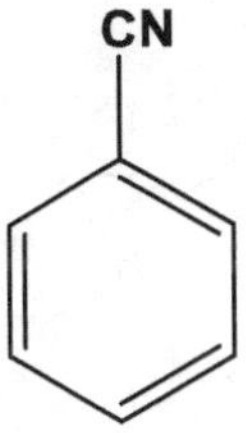


5

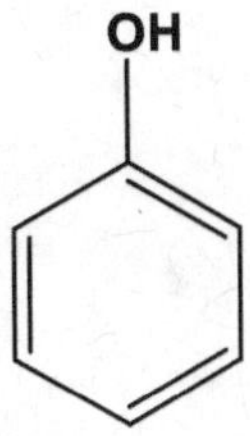


6

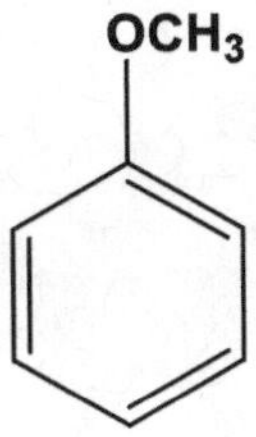


7

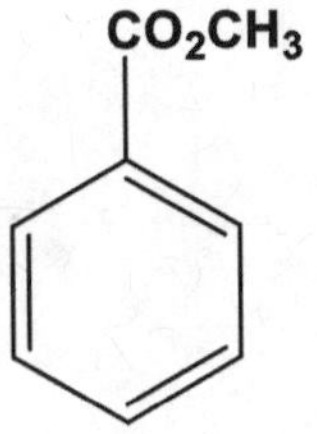


8

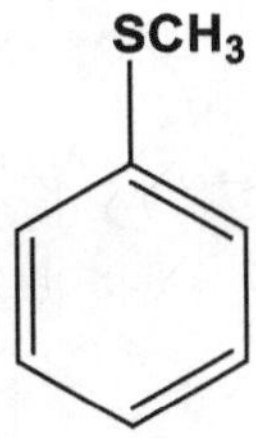


9

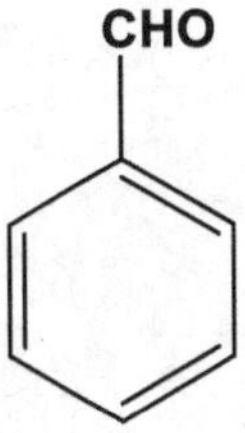


10

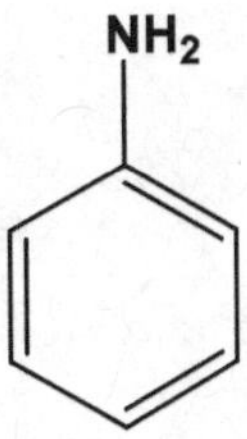


11

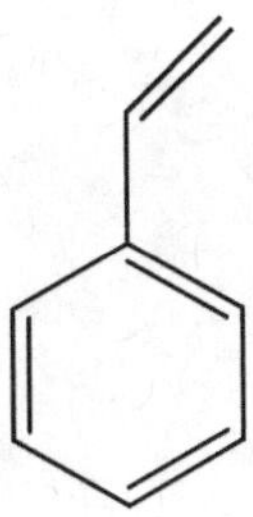

12

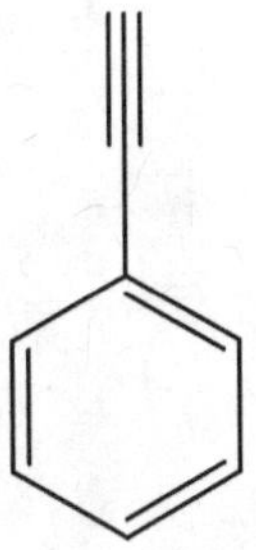

13

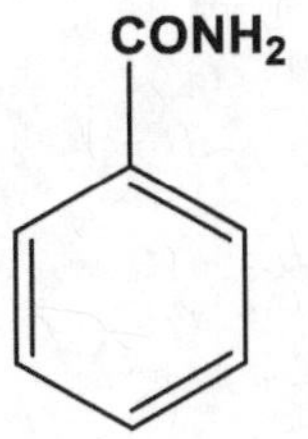


14

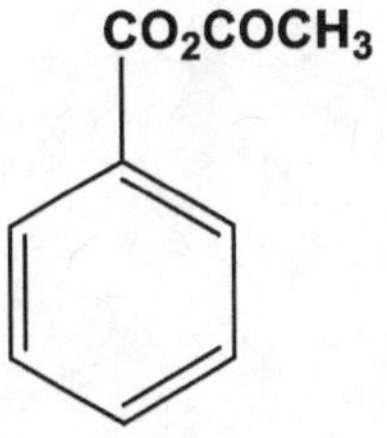


15

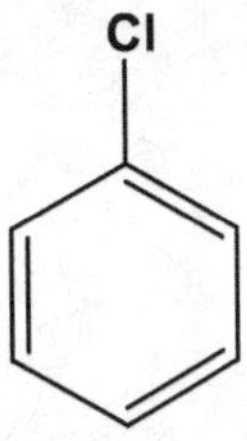


16

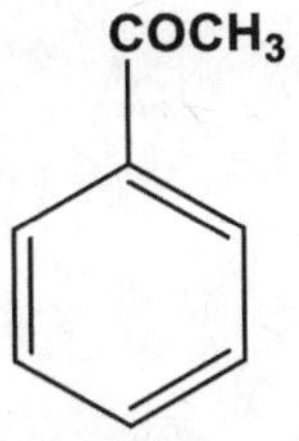


17

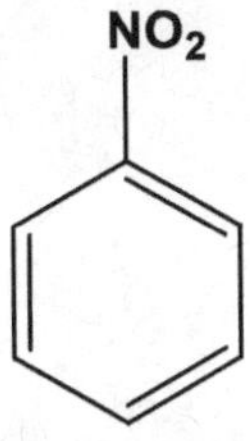


18

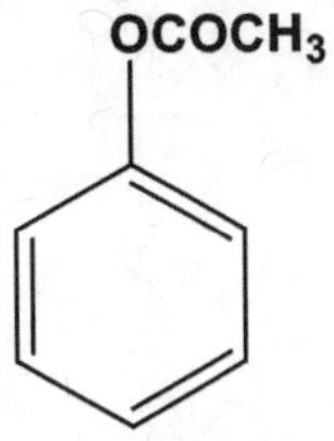


19

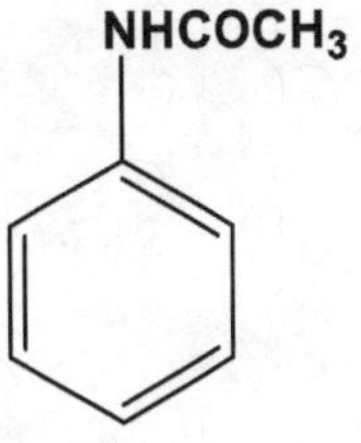


20

# 9

# ANSWERS

***For best results, avoid consulting the answers until AFTER trying the problems and thinking about the concepts carefully.***

## PROBLEM #1: LEWIS STRUCTURES AND RESONANCE

For:

$$H_3CCO_2CH_3 = H_3C\overset{O}{C}OCH_3$$

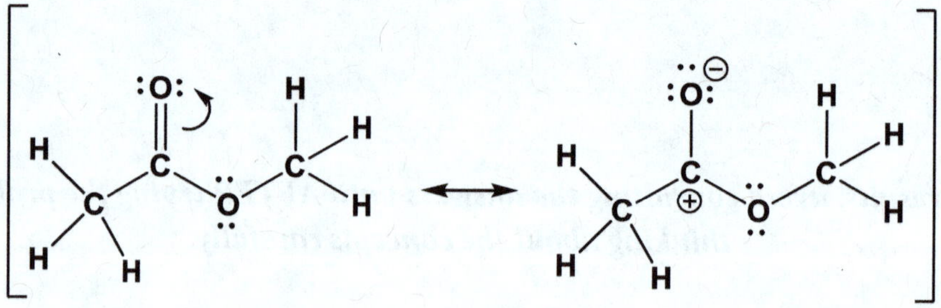

the structure on the left is the better contributor, as all atoms have an octet of electrons (**NOTE**: even though the structure on the right has formal charges properly localized and separated)

For:

$$H_3C\overset{O}{S}CH_3$$

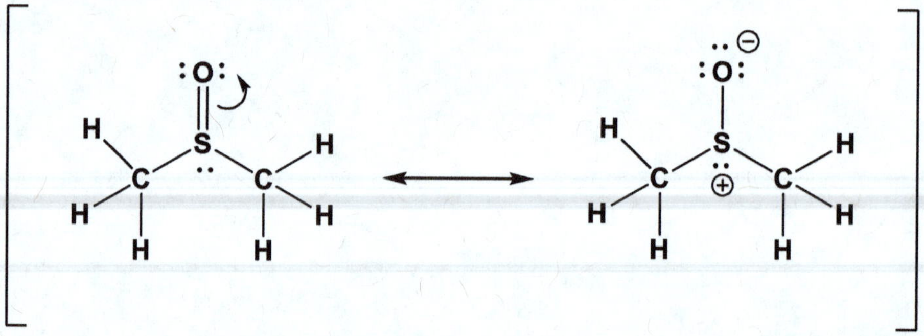

the structure on the right is the better contributor as all atoms have an octet of electrons (**NOTE**: even though the structure on the right has formal charges properly localized and separated)

NOTE: Although sulfur can expand its octet (making the structure on the left valid), ALL atoms prefer an octet. This preference is MUCH MORE important than the presence or absence of formal charges!

For:

$H_3CNO_2$

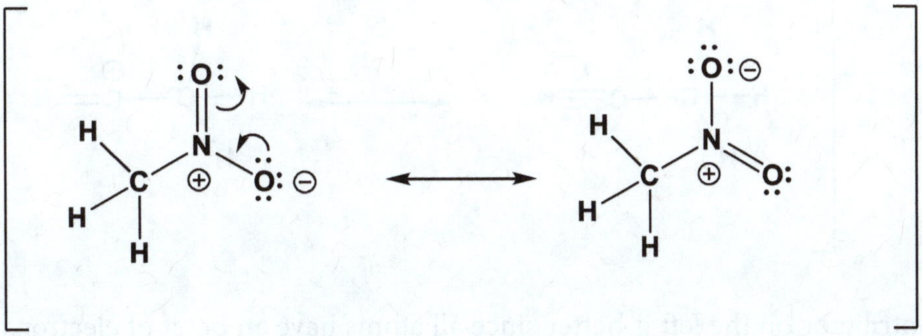

both structures contribute equally; they are "degenerate"

For:

$H_3CCOCH_3 = H_3C\overset{O}{C}CH_3$

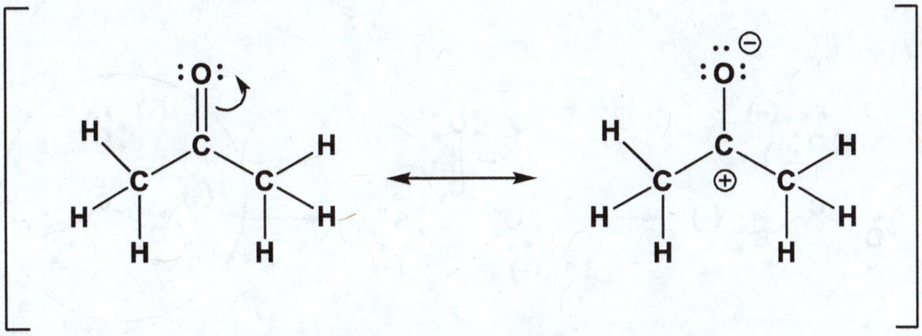

the contributor on the left is better since all atoms have an octet of electrons (**NOTE**: even though the structure on the right has formal charges properly localized and separated)

For:

**$H_3CCN$**

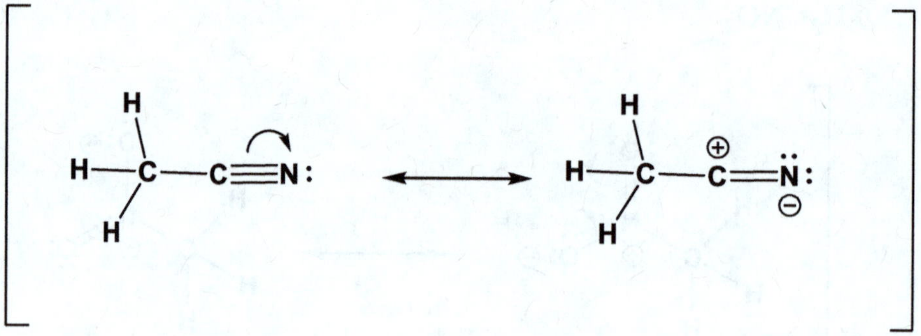

the contributor on the left is better since all atoms have an octet of electrons (**NOTE**: even though the structure on the right has formal charges properly localized and separated)

## PROBLEM #2: LEWIS STRUCTURES AND RESONANCE

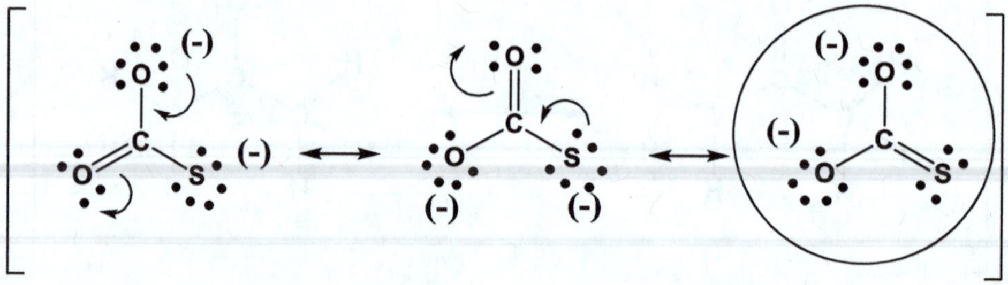

Reason the circled contributor would be best:

<u>**CHARGES APPROPRIATELY LOCALIZED**</u>

## PROBLEM #3: UDED: ASSESSING STABILITIES AND REACTIVITIES VIA RESONANCE

ANSWER:

(1) Stability: 1>4>2>3

(2) Reactivity: 3>2>4>1

Nucleophiles

## PROBLEM #4: UDED: ASSESSING STABILITIES AND REACTIVITIES VIA INDUCTION

ANSWER:

(1) Stability: 4>1>3>2

(2) Reactivity: 2>3>1>4

Nucleophiles

## PROBLEM #5: UDED: ASSESSING STABILITIES AND REACTIVITIES VIA HYBRIDIZATION

ANSWER:

(1) Stability: 3>1>2>4

(2) Reactivity: 4>2>1>3

Nucleophiles

## PROBLEM #6: REACTIONS OF NUCLEOPHILES

In the resonance example provided to you during the UDED recitation, you identified the 4 molecules as nucleophiles. This means they will react with electrophiles. For each molecule, show AT WHICH ATOM an electrophile will react. How many different products can arise from each nucleophile?

**ANSWER: (1): 2 products (2): 2 products (3): 1 product (4): 2 products**

In the induction example provided to you during the UDED recitation, you identified the 4 molecules as nucleophiles. This means they will react with electrophiles. For each molecule, show AT WHICH ATOM an electrophile will react. How many different products can arise from each nucleophile?

**ANSWER: 1 product each**

In the hybridization example provided to you during the UDED recitation, you identified the 4 molecules as nucleophiles. This means they will react with electrophiles. For each molecule, show AT WHICH ATOM an electrophile will react. How many different products can arise from each nucleophile?

**ANSWER: 1 product for 1-3; 2 products for (4)**

## PROBLEM #7: MORE RANKINGS BASED ON UDED

**Rank the following according to Decreasing Stability and Decreasing Reactivity**

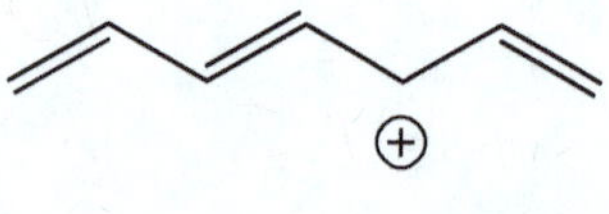

1

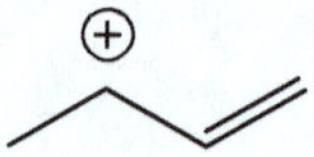

2

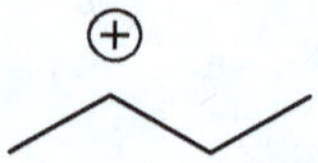

3

4

**Answer:** **Stability = E** **Reactivity = P**

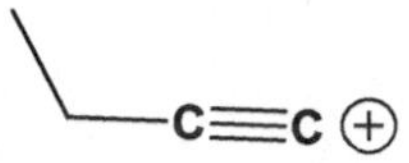


1

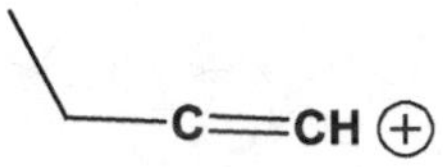


2

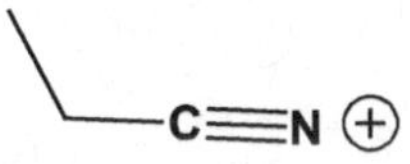


3

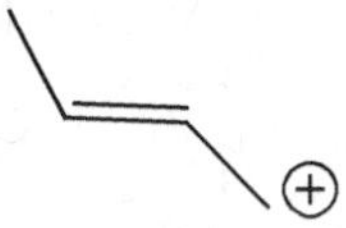

4

**Answer:** **Stability = W** **Reactivity = M**

**Note: all have resonance forms but #4 has the best resonance contributors...the others have equal number of poor contributors so #4 is most stabe. Next we considerhybridization and then C vs N electronegativity.**

**To consider hybridization, by our "rules", the carbon bearing the positive charge is sp for compound 2 and sp2 for compound 4. What about 1 and 3? In these cases the atom bearing the positive charge is unhybridized and the charge resides in an unhybridized p orbitatal attached to a carbon forming the sigma bond with an s-orbital**

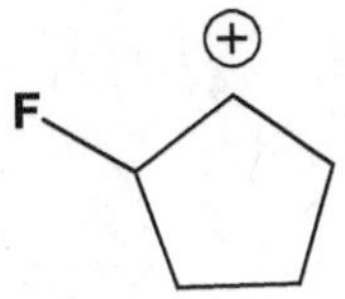


1

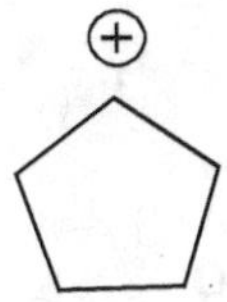

2

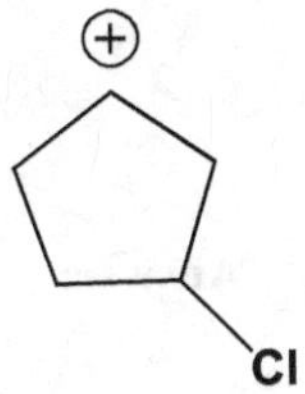


3

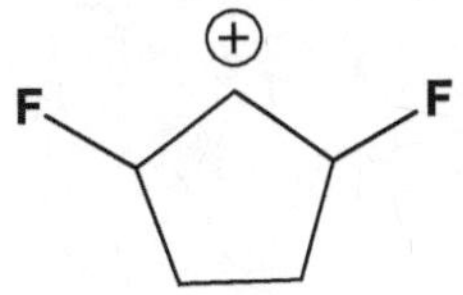


4

**Answer:** **Stability = I** **Reactivity = T**

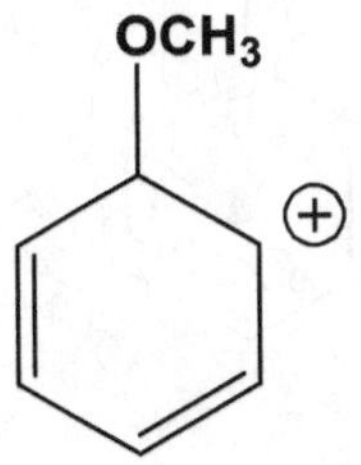


1

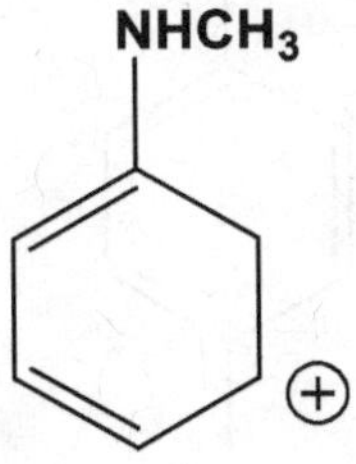


2

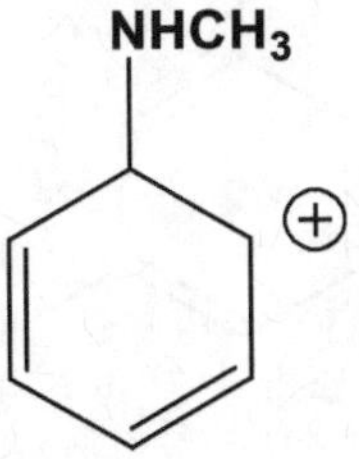


3

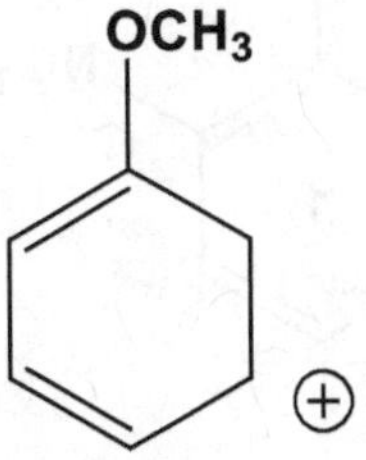


4

**Answer:** **Stability = L** **Reactivity = D**

Rank the following in terms of Decreasing Basisity

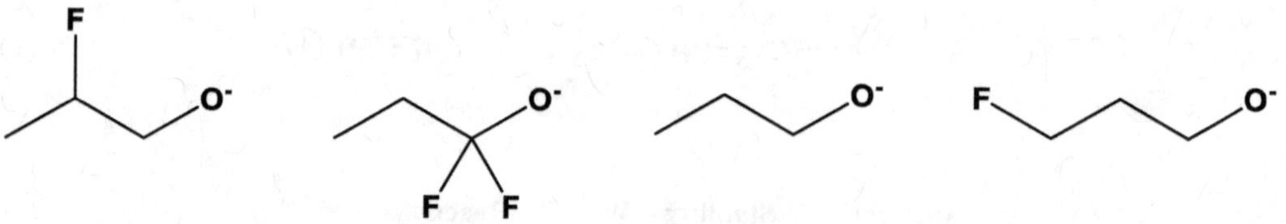


1 2 3 4

**Answer:** **Basisity = Q**

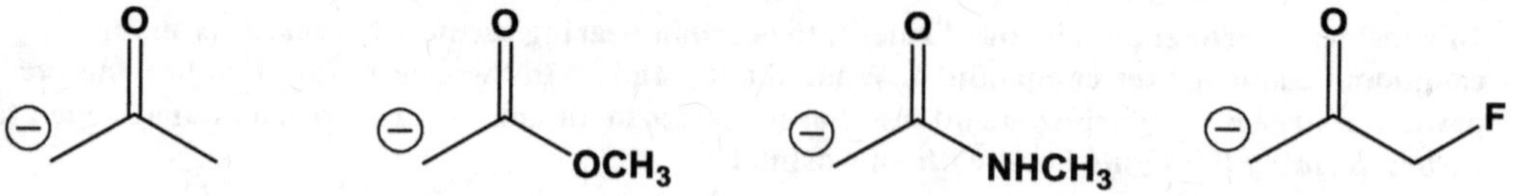


1 2 3 4

**Answer:** **Basisity = O**

Rank the following in terms of Decreasing Acidity

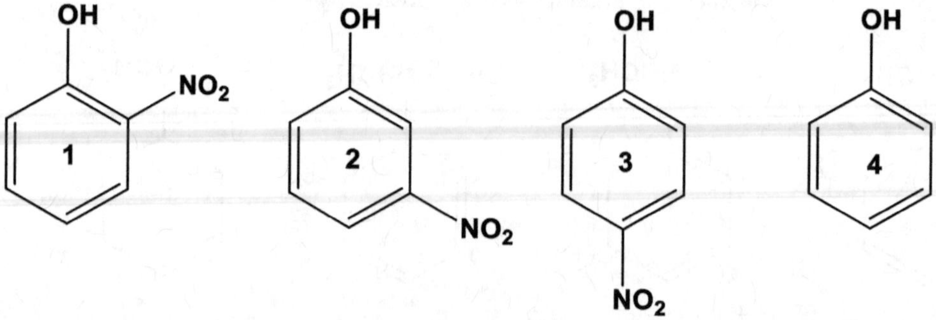


**Answer:** **Acidity = C**

## PROBLEM #8: UDED-ELECTRON DONATING AND WITHDRAWING

(a)

1. alkane
2. carboxylic acid
3. thiol
4. acyl halide (chloride)
5. nitrile
6. alcohol
7. ether
8. ester
9. sulfide (not on table 2-3)
10. aldehyde
11. amine
12. alkene
13. alkyne
14. amide
15. amhydride
16. halide (chloride)
17. ketone
18. nitro (not on table 2-3)
19. ester
20. amide

(b): D = "donate" W = "withdraw N/A = not applicable

note: Although in principle, carbonyl groups can withdraw or donate electron density via resonance, the high electronegativity of oxygen makes electron withdrawal via resonance the predominant UDED resonance behavior of this moiety

1. N/A
2. W
3. D
4. W
5. W
6. D

7. D
8. W
9. D
10. W
11. D
12. D or W (can do either)
13. D or W (can do either)
14. W
15. W
16. D
17. W
18. W
19. D
20. D

(c)

1. D (you will not learn this until Chapter 3)
2. W
3. W
4. W
5. W
6. W
7. W
8. W
9. W
10. W
11. W
12. W (this is more of a hybridization effect)
13. W (this is more of a hybridization effect)
14. W
15. W
16. W
17. W
18. W

19. W
20. W

(d) Resonance is a dominant UDED effect when compared to induction. There are unique situations where induction can overcome resonance, especially when induction is strong and resonance is poor. These exceptions will always be pointed out to you. You are not expected to know them until I mention them specifically in lecture.

(e) 2; 4; 8; 10; 14; 15; 17; 19; 20

(f) NO: examine esters 8 and 19 and amides 14 and 20...remember that resonance is the dominant UDED feature

(g)

20>14>8>17

19>8>15>17

www.ingramcontent.com/pod-product-compliance
Lightning Source LLC
LaVergne TN
LVHW060453061126
851857LV00016B/1815